IRREGULAR SHAPE ANCHOR IN COHESIONLESS SOILS

IRREGULAR SHAPE ANCHOR IN COHESIONLESS SOILS

HAMED NIROUMAND

Butterworth-Heinemann
An imprint of Elsevier

Butterworth-Heinemann is an imprint of Elsevier
The Boulevard, Langford Lane, Kidlington, Oxford OX5 1GB, United Kingdom
50 Hampshire Street, 5th Floor, Cambridge, MA 02139, United States

Library of Congress Cataloging-in-Publication Data
A catalog record for this book is available from the Library of Congress

British Library Cataloguing-in-Publication Data
A catalogue record for this book is available from the British Library

ISBN: 978-0-12-809550-8

For information on all Butterworth-Heinemann publications
visit our website at https://www.elsevier.com/books-and-journals

Working together
to grow libraries in
developing countries

www.elsevier.com • www.bookaid.org

Publisher: Joe Hayton
Acquisition Editor: Andre Gerhard Wolff
Editorial Project Manager: Jennifer Pierce
Production Project Manager: Kiruthika Govindaraju
Cover Designer: Alan Studholme

Typeset by SPi Global, India

DEDICATION

Dedicated to my mom

CONTENTS

LIST OF TABLES

LIST OF FIGURES

ABOUT THE AUTHOR

Hamed Niroumand is an assistant professor at the Department of Civil Engineering, Buein Zahra Technical University. He is currently the Vice Chancellor for the Research and Academic section of Buein Zahra Technical University. His main fields of research are geotechnical engineering, earth anchors, deep foundation, numerical analysis, sustainable development, and nano-materials. He has been a project manager and professional engineer in various geotechnical and earth-building projects. Between 2011 and 2015 he received four medals and international awards for his inventions and research, including first place for research at the national Iranian young inventor and researcher festival 2012 and first place for research at the national Iranian youth festival in 2012 and 2013. In 2016, he received the best researcher award from the Ministry of Road and Urban Development. He has been the chairman and head director of international/national conferences on civil engineering for close to 20 events held in various countries. Niroumand has chaired sessions at several international/national conferences and festivals in various countries and has presented various research papers in many conferences around the world. He has published approximately 200 papers in journals and conferences. He is also an editorial team member and reviewer for scientific journals. He has about 15 inventions credited to him, which are patented/patent pending.

Personal website: www.niroomand.net

NOTATIONS

A	Area of anchor plate
B	Length of anchor plate/irregular shape anchor
C_u	Coefficient of uniformity
D	Width of anchor plate/irregular shape anchor
D_{10}	Effective grain size
D_{50}	Uniformity of sand
e	Void ratio
G_s	Specify gravity
H	Depth of soil
H_{cr}	Critical embedment ratio
I_d	Density index
K_o	Earth pressure coefficient at rest
K_u	Nominal pullout coefficient of earth pressure on a convex cylindrical
L	Depth of soil above anchor plate/irregular shape anchor
L/D	Embedment ratio
M	Mass of soil
n	Porosity
N_q	Breakout factor
n_r	Relative porosity
$\emptyset$	Soil friction angle
P	Pullout load
P_s	Shearing resistance
P_t	Force below of area
P_u	Ultimate pullout load
S_f	Shape factor
t	Time
V	Volume of soil
W	Effective weight of soil
γ	Unit weight of soil
γ_d	Dry unit weight of soil

LIST OF APPENDICES

CHAPTER 1

Introduction

1.1 INTRODUCTION

Structures such as transmission towers, tunnels, sea walls, buried pipelines, retaining walls and others are subjected to considerable pullout forces. In such cases, an economic design solution may be obtained through the use of tension members. These elements, which are related to anchors, are generally fixed to the structure and embedded in the ground to effective depth so that they can resist uplifting forces.

Most of the anchoring systems used in various countries are a grout-base anchoring system. The installation of these may take several days. The *irregular shape anchor* is a new anchoring system that can be installed in sand without having to be grouted. This system is expected to increase both the speed of construction and cost effectiveness. Reviews of previous researchers' activities are presented in Chapter 2, including analytical procedures. This chapter is a review based on an introduction to the topic of plate anchors, because the irregular shape anchor is a state-of-the-art product in plate anchor design.

Field testing and two model laboratory tests are used to research the performance of a horizontal irregular shape anchor subjected to pullout loading. Field studies in plate anchors, reported by earlier researchers, are discussed in Chapter 2.

Review of theories and numerical analysis of plate anchors from previous researchers are discussed in Chapter 2, including the analytical procedures. Analyses, beginning with Meyerhof and Adams (1968) until the most recent analyses, such as Kuzer and Kumar (2009), are also reviewed.

Research methodologies are discussed in Chapter 3. The irregular shape anchor tests were performed to research pullout tests in loose and dense sand around the anchor in a chamber box, in which the aforementioned system rotates after the beginning of pullout loading. Raining methods and the vibration method were employed in the chamber box to obtain loose and dense packing sand, respectively.

Chapter 4 contains the interpretation of the experimental results. The variations of model ultimate capacity P_u with the embedment ratio L/D

Irregular Shape Anchor in Cohesionless Soils
http://dx.doi.org/10.1016/B978-0-12-809550-8.00001-0

and N_q breakout factor in dry loose sand and dense sand are analyzed. Dry sand was obtained from a Malaysia sand quarry area, generally having grain sizes between 0.205 and 2.36 mm. This was used as an embedment medium in both small and large models in the wood box. Due to the difficulties of preparing similar sand sizes, a wide range of size between 0.07 and 5.01 mm was used for the embedment of small and large model sizes of the irregular shape anchor in the wood box. It was assumed that no prominence effect on the stress value would be encountered between these two arrangements. The irregular shape anchor blade, joint and rods were fabricated as a new anchor in a special shape, with lengths of 290 and 145 mm and widths of 100 and 50 mm, so that the anchors can be taken into sand and, after putting them into balanced embedment, the rotation begins.

Chapter 5 contains the interpretation of the theoretical analysis results. The new empirical formula of model ultimate capacity P_u with the embedment ratio L/D and N_q breakout factor in dry loose sand and dense sand were compared to previous theories of anchor plates. The proposed results are compared with previous works, such as Murray, Meyerhof and Dickin.

Chapter "Conclusions" outlines the main conclusions of this research and offers suggestions for future work.

CHAPTER 2

Literature Review

2.1 INTRODUCTION

In this chapter, a review is provided of previous theoretical and experimental work, and the analysis of an anchor plate embedded in sand experiencing pullout loading is considered.

Researchers such as Mors (1959), Giffels et al. (1960), Balla (1961), Turner (1962), Ireland (1963), Sutherland (1965), Mariupolskii (1965), Kananyan (1966), Baker and Konder (1966), Adams and Hayes (1967), Mors (1959), Balla (1961), Andreadis et al. (1981), Dickin (1988), Frydman and Shamam (1989), Ramesh Babu (1998), Krishna (2000), Fargic and Marovic (2003), Merifield and Sloan (2006), Dickin and Laman (2007), Kumar and Kouzer (2008), and Kuzer and Kumar (2009) have been especially concerned with the general solution for an ultimate pullout capacity in sand. One of the applications of soil anchors has been in transmission towers. This application was preliminary for much of the initial research into anchor behavior (Balla, 1961). Initially these towers were supported by large dead-weight concrete blocks, where the required uplift capacity was obtained solely due to its own weight. This simple design came at considerable cost and, as a result, research was undertaken to find a more inexpensive design solution. As the range of applications for anchors expanded to include the support of different structures, a more concerted research effort has meant that soil anchors today have evolved to the point where they now provide an inexpensive and competitive alternative to these mass foundations. Generally, research into the behavior of anchors can take one of two forms: experimental or numerical/theoretical-based studies. In this chapter, no attempt has been made to present complete information on all research performed, but rather a more selective summary of research is presented. For example, although theory portions of the underlying capacity of grouted anchors are also applicable to typical soil pullout behavior, our discussion is limited only to the current proposal topic as related to plates. Thus, systems related to the behavior of helical anchors have not been reviewed. It will

Irregular Shape Anchor in Cohesionless Soils
http://dx.doi.org/10.1016/B978-0-12-809550-8.00002-2

become clear that past research has been experimentally based and, as a result, current design practices are largely intuitive. In contrast, very few thorough numerical analyses have been performed to determine the ultimate pullout loads of different systems. Also, of the numerical studies that have been published in the literature, few can be considered to be accurate (Figs. 2.1–2.7).

2.2 REVIEW OF PREVIOUS EXPERIMENTAL WORKS

Numerical analysis and laboratory works have been used to research the performance of the ultimate pullout loading anchor plate. However, there are relatively few brief papers in the technical literature that deal with an anchor

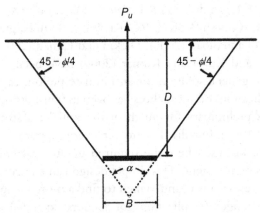

Fig. 2.1 Failure surface assumed by Mors (1959).

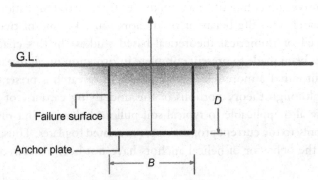

Fig. 2.2 Failure surface assumed by Downs and Chieurzzi (1966).

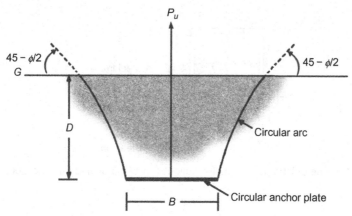

Fig. 2.3 Failure surface assumed by Balla (1961).

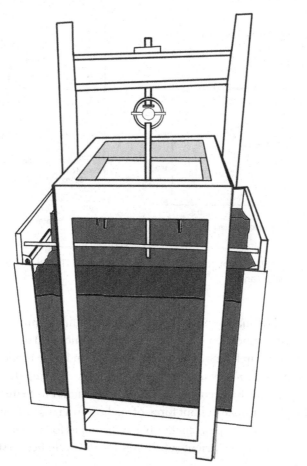

Fig. 2.4 Laboratory test devices of Fargic and Marovic (2003).

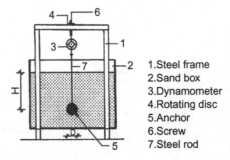

1. Steel frame
2. Sand box
3. Dynamometer
4. Rotating disc
5. Anchor
6. Screw
7. Steel rod

Fig. 2.5 Scheme of laboratory tests for pullout test of Fargic and Marovic (2003).

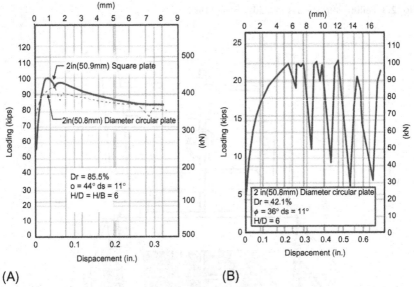

(A)   (B)

Fig. 2.6 Load/displacement curves of Murray and Geddes (1987): (A) Very dense sand; (B) medium dense sand.

plate in sand. For analysis purposes, the limiting ultimate pullout capacity of an anchor plate embedded in sand can be carried out based on numerical and experimental research. Laboratory tests on small and large models have been performed in order to obtain appropriate solutions. During the last 50 years various researchers have conducted laboratory studies to better understand and predict the ultimate pullout force of anchors in a range of soil types. The previous data has focused on anchor behavior and its force in sand. Although they are not entirely sufficient substitutes for full-scale field testing, tests at

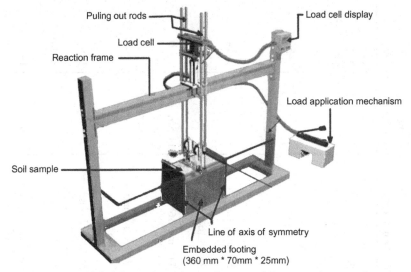

Fig. 2.7 3D view of the chosen experimental set-up of Kumar and Kouzer (2008).

laboratory scale have the advantage of allowing close control of at least some of the parameters encountered in research. In this way, behavior shown in the laboratory can be of value in developing an understanding of performance at larger scales. In addition, observations made in laboratory testing can be used in conjunction with numerical analysis to develop different theories. These solutions can then be applied to solve a wider range of problems (Table 2.1).

Experimental investigations into plate anchor behavior have typically adopted one of two selections: namely, conventional methods under "normal gravity" conditions, or centrifuge systems. Centrifuge systems use physical scaling laws to match the model and prototype behavior and can be used to research mechanical anchors. These investigations are based on soil stress fields that are in proportion to the size of the model plates. In this way, a particular anchor size buried at a sufficient depth can also be used to investigate a range of burial depths simply by verifying the stress field. While at rest, the set-up is subject to a static gravitational force equal to 1.00 g. By keying the model in a centrifuge motion, gravitational forces greater than 1.00 g can be obtained, which makes the required stress field and, in turn, simulates in situ stresses for different burial depths. Unfortunately, because of the basic equipment and set-up costs, only a few documented institutions have such a tool at their disposal.

Table 2.1 Last experimental model tests on horizontal anchors in sand

Anchor	Type of testing	Anchor shape	Anchor size	Friction angle	Anchor roughness	L/D
Balla (1961)	Laboratory	Circular	–	Dense 37°	–	1–5
Hanna and Carr (1971)	Laboratory	Circular	38mm	37°	–	4–11.2
Hanna et al. (1971)	Laboratory and field	Circular	38 mm and 150 mm	37°	–	4–11.2
Das and Seeley (1975a,b)	Laboratory	Square rectangular	51 mm	31°	–	1–5
Rowe (1978)	Laboratory	Square rectangular	51 mm	32°	16.7°	1–8
Andreadis et al. (1981)	Laboratory	Circular	80 mm–150 mm	37° and 42.5°	–	1–14
Ovesen (1981)	Centrifuge and field	Circular square	20 mm	29.5° and 37.7°	–	1–3.39
Murray and Geddes (1987)	Laboratory	Circular rectangular	50.8 mm	44° and 36°	11 Smooth and 42 rough	1–10
Frydman and Shamam (1989)	Laboratory and field	Strip rectangular	19 mm and 200 mm	30° Loose and 45° dense	–	2.5–9.35
Dickin (1988)	Centrifuge and laboratory	Square rectangular	25 mm and 50 mm	38°–41° Loose and 48°–51° dense	–	1–8
Tagaya et al. (1988)	Centrifuge	Circular and rectangular	15 mm	42°	–	3–7.02
Murray and Geddes (1987)	Laboratory	Square rectangular	50.8 mm	36° Loose and 43.6° dense	10.6°	1–8

Reference		Shape	Size	Friction angle		
Sarac (1989)	—	Circular and square	—	37.5°, 48°	—	0.35–4
Bouazza and Finlay (1990)	Laboratory	Circular	37.5 mm	33.8°, 39°, 4 3.7°	—	2–5
Dickin (1994)	Centrifuge	Strip and pipe	25 mm	38°–41° Loose and 48°–51° dense	—	2–7
Sakai and Tanaka (1998)	Laboratory	Circular	30 mm–200 mm	—	—	1–3
Ramesh Babu (1998)	Laboratory	Square, circular and strip	50 mm, 75 mm, 100 mm	32.5° Loose and 45° dense	—	1–8
Pearce (2000)	Laboratory	Circular	50 mm–125 mm	Loose to very dense	—	2–15
Fargic and Marovic (2003)	Laboratory and field	Spatial	25 mm, 50 mm and 100 mm	Loose and dense	—	1–5
Murray and Geddes (1987)	Laboratory	Circular	50.8 mm–88.9 mm	36° Medium and 44° dense	—	1–10
Dickin and Laman (2007)	Centrifuge	Strip	100 mm–250 mm	35° Loose and 51° dense	—	1–8
Kumar and Kouzer (2008)	Laboratory	Multiple strip	70 mm–370 mm	37.4°, 41.8°, 44.8°	—	0.41–12.8 6

In comparison to centrifuge testing, the more conventional gravity method is typically a cheaper testing alternative, due to the ease of set-up and the need for only basic equipment. Quite often, a conventional gravity method can be incorporated into a geotechnical engineering research laboratory by use of existing equipment. Unfortunately, full-scale testing of foundations is very expensive, time consuming and in most cases difficult and rough. For these reasons, testing is typically limited to small-scale model tests, as these provide an economical and convenient alternative. These are most commonly used for sands, and centrifuge testing is also used. Of course, both methods have advantages and disadvantages associated with them, and these must be kept in mind when interpreting the results from experimental studies of anchor behavior. The following sections provide a brief abstract of early and recent experimental research into the behavior of plate anchors in different soils.

2.2.1 Methods of Analysis in Anchor Plate

A failure mechanism was adopted and the pullout force was then determined by considering the equilibrium of the soil mass above the anchor plate and included by the adopted failure surface. Based on the underlying adoption, these methods of analysis can be separated into two types:

- The "soil cone" method (Mors, 1959) in which the failure surface consists of a small cone extending from the anchor edges up to and intersecting the soil surface at an angle of $\left(45 - \dfrac{\phi'}{2}\right)°$. The anchor force is defaulted to equal the weight of the soil contained within the area of the assumed failure surface. Any shearing resistances and their force developed along the failure surface are very small.
- The "friction cylinder" method (Downs and Chieurzzi, 1966), in which failure is assumed to occur along the surface of a cylinder of soil above the anchor. The anchor capacity is defaulted to equal the sum of the weight of the soil contained within the area of the default cylinder to failure surface, and the frictional resistance and their force derived along the failure surface.

2.2.2 Previous Experimental Works

Subsequent variations on these early theories have been proposed, including that of Balla (1961) who determined the shape of slip surfaces for horizontal shallow anchors in dense sand and proposed a numerical method for estimating the force of anchors based on the observed shapes of the slip surfaces.

Baker and Konder (1966) approved Balla's findings regarding the behavioral difference of deep and shallow anchors in dense sand (Table 2.2).

Sutherland (1965) showed results for the pull-out of 150 mm horizontal anchors in loose and dense sand, as well as large diameter shafts in medium dense to dense sands. It was concluded that the mode of failure is different with sand density and Balla's analytical approach may give reasonable results only in sands of intermediate density. Kananyan (1966) showed results for horizontal circular plate anchors in loose to medium dense sand. He also performed a series of tests on inclined anchors and studied the failure surface, concluding that most of the soil particles above the anchor moved predominantly in a vertical direction. In these tests, the ultimate pullout force increased with the inclination angle of the anchors.

Das and Seeley (1975a,b) performed uplift tests for horizontal rectangular anchors $(L/D \leq 5)$ in dry sand with a friction angle of $\varphi = 31$ degrees at a density of 14.8 kN/m^3. For each aspect ratio (L/D), the anchor capacity was found to increase with the embedment ratio before reaching a constant value at the critical embedment depth. A similar investigation was conducted by Rowe (1978) in dry sand with friction angles between 31 degrees and 33 degrees, and dry unit weight of $\gamma = 14.9$ kN/m^3. Polished steel plates were used for the anchors and the interface roughness was measured as $\delta = 16.7$ degrees. Most tests were conducted on anchors with an aspect ratio L/D of 8.75. This suggests that anchors with aspect ratios of $L/D > 5$ effectively behave as a continuous strip and can be compared with methods that obtain plane strain conditions. In contrast to the observations of Das and Seeley (1975a), Rowe (1978) did not observe a critical embedment depth and the anchor capacity was found to continually increase with embedment ratios over the range of $L/D = 1$–8.

Extensive chamber testing programs have been conducted by Murray and Geddes (1987), who performed pullout loading tests on horizontal strip, circular, and rectangular anchor plates in dense and medium dense sand with $\varphi = 43.6$ degrees and $\varphi = 36$ degrees, respectively. Anchors were typically 50.8 mm in width/diameter and were tested at aspect ratios (L/D) of 1, 2, 5, and 10. Based on the results seen, Murray and Geddes came to several conclusions:

1. The pullout loading force of rectangular anchor plates in very dense sand increases with embedment ratio and with decreasing aspect ratio L/B.
2. There is a role difference between the force of horizontal anchors with rough surfaces compared to those with area smooth surfaces (as much as 15%).

Table 2.2 Previous theoretical analysis on horizontal anchors in sand

Researcher	Analysis method	Anchor shape	Anchor roughness	Friction angle	L/D
Meyerhof and Adams (1968)	Limit equilibrium	Strip, square and circular	—	—	—
Vesic (1971)	Cavity expansion	Strip and circular	—	0–50°	0–5
Rowe and Davis (1982)	Elastoplastic finite element	Strip	Smooth	0–45°	1–8
Vermeer and Sutjiadi (1985)	Elastoplastic finite element	Strip	—	All	1–8
Tagaya et al. (1983, 1988)	Elastoplastic finite element	Circular and rectangular	—	31.6°, 35.1°, 42°	0–30
Saeedy (1987)	Limit equilibrium	Circular	—	20°–45°	1–10
Murray and Geddes (1987)	Limit analysis and limit equilibrium	Strip, square and circular	—	All	All
Koutsabeloulis and Griffiths (1989)	Finite element method–initial stress	Strip and circular	—	20°, 30°, 40°	1–8
Sarac (1989)	Limit equilibrium	Circular and square	—	0–50°	1–4
Basudhar and Singh (1994)	Limit analysis	Strip	Rough/ smooth	32°	1–8
Kanakapura et al. (1994)	Method of characteristics	Strip	Smooth	5°–50°	2–10
Ghaly and Hanna (1994a)	Limit equilibrium	Circular	—	30°–46°	1–10
Smith (1998)	Limit analysis	Strip	—	25°–50°	1–28

Sakai and Tanaka (1998)	Elastoplastic finite element	Circular	—	Dense	1–3
Krishna (2000)	Finite difference method	Strip	—	30° Loose and 45° dense	1–8
Fargic and Marovic (2003)	Elastoplastic finite element	Spatial	—	Loose and dense	1–5
Merifield and Sloan (2006)	Limit analysis	Circular and square	—	20°–40°	1–10
Dickin and Laman (2007)	Finite element method	Strip	—	35° Loose and 51° Dense	1–8
Kumar and Kouzer (2008)	Finite element method	Multiple strip	—	37.4°, 41.8°, 44.8°	0.41–12.86
Kuzer and Kumar (2009)	Limit analysis and displacement finite element method	Strip	Rough	25°, 30°, 35° and 45°	0–5.03

3. Experimental results suggest that an anchor plate with an aspect ratio of $L/D=10$ behaves like a strip and does not differ much from an anchor with $L/D=5$.
4. The force of circular anchor plates in very dense sand is approximately 1.26 times the capacity of square anchors.

The validation of these conclusions confirms the findings of Rowe (1978). It is also of interest to note that for all the tests performed by Murray and Geddes, no critical embedment depth was seen.

More recently, Pearce (2000) performed a series of laboratory pullout tests on horizontal circular plate anchors pulled vertically in dense sand. These tests were conducted in a large chamber box, with dimensions of 1 m in height and 1 m in diameter.

Various parameters such as anchor plate diameter, pullout rate and elasticity of the loading system have been investigated. The model anchor plates used for the pullout tests varied in diameter between 50 and 125 mm and were constructed from 28 mm thick mild steel. Large diameter anchors were selected (compared with previous research) due to the identified influence of scale effects on the break-out factor for anchors of diameters less than 50 mm (Andreadis et al., 1981). Dickin (1988) performed 41 tests on 25 mm anchor plates with aspect ratios of $L/D=1-8$ at embedment ratios L/D up to 8 in both loose and dense sand. A number of conventional gravity tests were also performed and compared to the centrifuge data. This comparison revealed a significant difference between the prediction anchor plate forces, particularly for square anchors where the conventional test output gave anchor forces up to twice those given by the centrifuge. Without explaining why, Dickin concluded that direct extrapolation of conventional chamber box test results to field scale would provide overly optimistic predictions of the ultimate force for rectangular anchor plates in sand. Tagaya et al. (1988) also performed centrifuge testing on rectangular and circular anchor plates, although the research was limited in comparison to that of Dickin (1988), discussed previously.

Dickin (1988) studied the influence of anchor geometry, embedment depth and the soil density on the pullout capacity of 1-m prototype anchor plate by subjecting 25-mm models to an acceleration of 40 g in a Liverpool centrifuge. It was found that for strip anchors, pullout resistance expressed as dimensionless breakout factor increases significantly with anchor embedment depth and soil density but reduces with an increase in value of embedment ratio, which is the ratio of length to width of the strip anchor plate. Failure displacements also increase with embedment depth but reduce with soil density and aspect value ratio.

Frydman and Shamam (1989) performed a series of pullout tests on prototype slabs placed at various inclinations and different depths in dense sand. A simple semiempirical expression is found to reasonably predict the pullout capacity of a continuous, horizontal slab as a foundation of depth-to-width ratio in their tests. Factors to account for shape and inclination are then made, leading to expressions for the estimation of pullout capacity of any slab anchor.

The following expressions have been proposed for the pullout capacity of a horizontal, rectangular slab anchor plate in sand, for dense sand:

$$(N_q)_r = \left[1 + \frac{D}{B}\tan\varphi\right]\left[1 + \frac{\left(\frac{B}{L} - 0.15\right)}{(1 - 0.15)} \times \left(0.51 + 2.35\log\left(\frac{D}{B}\right)\right)\right]$$

(2.1)

For loose sand, $D/B \geq 2$:

$$(N_q)_r = \left[1 + \frac{D}{B}\tan\varphi\right]\left[1 + 0.5\frac{\left(\frac{B}{L} - 0.15\right)}{(1 - 0.15)}\right]$$

(2.2)

where $(N_q)_r$ is the pullout capacity.

Ramesh Babu (1998) investigated the pullout capacity and the load deformation behavior of a shallow horizontal anchor plate. Laboratory experiments have been conducted on anchors of different shapes (square, circular and strip) and embedded in medium dense and dense sands. In addition, the effect of submergence of soil above anchor plates has been investigated. Ashraf Ghaly had recommended a general expression for pullout capacity of vertical anchor plates based on static analysis of the experimental test results from the published literature. Along similar lines and incorporating appropriate corrections, Ramesh Babu (1998) proposed a general expression for horizontal anchor plates in sand by analyzing the results of published experimental data and his own pullout test data.

For a horizontal strip anchor plate:

$$\frac{P_u}{\gamma A D \tan\varphi} = 3.24\left(\frac{D^2}{A}\right)^{0.34}$$

(2.3)

For a square and a circular anchor plate:

$$\frac{P_u}{\gamma AD \tan \varphi} = 3.74 \left(\frac{D^2}{A} \right)^{0.34} \tag{2.4}$$

where $\dfrac{P_u}{\gamma AD \tan \varphi}$ is pullout capacity factor and D^2/A is a geometry factor.

2.2.3 Recent Experimental Work

Only a few investigations into the performance of ultimate pullout loading in cohesionless soil were recorded in model laboratory studies. One example is the work of Fargic and Marovic (2003), which discussed the pullout capacity of anchors in soil under applied uplift force. In field tests, the pullout forces were gradually increased and the earth surface displacements measured in two profiles perpendicular to each other. Laboratory and field tests were performed for several embedment depths and anchor plate diameter ratios in the same sand and under the same conditions.

Murray and Geddes (1987) investigated the vertical pullout of anchor plates in medium sand; factors investigated in relation to the load-displacement response were the size and shape of plate, depth of embedment, sand density and plate surface roughness in the laboratory. Significant differences in behavior were noted between anchor plates embedded in very dense sand and those embedded in medium dense sand.

Dickin and Laman (2007) investigated the physical modeling of anchor plates in the centrifuge. The centrifuge incorporates balanced swinging buckets, which were basically 0.57 m long, 0.46 m wide and 0.23 m deep. Physical research investigated the pullout response of 1-m wide strip anchors in sand; the results indicated maximum resistance increases with anchor embedment ratio and sand packing. The physical research showed that breakout factors for 1-m wide strip anchors increased with anchor embedment ratio and sand packing.

Kumar and Kouzer (2008) used a group of multiple strip anchors placed in sand and subjected to equal magnitudes of vertical upward uplift loads in model experiments. Instead of using a number of anchor plates in the experiments, a single anchor plate was used by modeling the boundary conditions along the plane of symmetry on both of the sides of the anchor plate. The effect of interference due to a number of multiple strip anchor plates placed in a granular medium at different embedment depths was investigated by conducting a series of small-scale model tests. The experimental data clearly reveals that the magnitude of the failure load reduces quite extensively with a decrease in the spacing between the anchor plates.

2.2.4 Limitations of Experimental Work

Different anchor plates and soil parameters were employed by different researchers. Inevitably, such a wide range of parameters will contribute to conflicting conclusions for ultimate pullout loading of anchor plates.

Some of the work did not include the internal friction angle, anchor roughness or anchor size, although most authors obtained their internal friction angle using the direct shear test or triaxial compression test.

Unfortunately, results obtained from laboratory testing are typically problem-specific and are difficult to extend to field problems with different material or geometric parameters in field scale.

2.3 PREVIEW OF PREVIOUS THEORETICAL AND NUMERICAL WORKS

A review of previous theoretical and numerical works and the analysis of ultimate pullout capacity of anchor plates in sand are considered in this chapter. Numerical analysis and laboratory works were performed to investigate the performance of ultimate pullout loading of anchor plates. However, there are relatively few documents in the technical literature that deal with an anchor plate in sand. For analysis purposes, the limiting ultimate pullout capacity of an anchor plate embedded in sand can be carried out based on numerical and experimental work. During the last 50 years various researchers have conducted laboratory studies to better understand and predict the ultimate uplift loading of anchors in a range of soil types. In contrast to the variety of experimental results already discussed, very few rigorous numerical analyses have been performed to determine the pullout loading of anchor plates in soil. In the literature, several theories have been proposed to calculate pullout force of anchors, the difference between each one of them being mainly in the shape of the selected failure surface. It is essential to verify theoretical solutions/numerical analysis with experimental studies wherever possible, as results assumed from laboratory testing alone are typically problem-specific. This is important in the case of geomechanics, where we are dealing with a highly nonlinear material that often displays pronounced scale effects. Thus, it is often difficult to extend the findings from the small scale of laboratory research to full-scale problems with different material or geometric parameters. Existing numerical analyses have generally selected a condition of plane strain for the case of a continuous strip anchor plate or axisymmetry for the case of circular anchors.

2.3.1 Numerical Methods

Numerical methods for estimating the ultimate pullout capacity of plate anchors have been developed. One of the earliest publications concerning ultimate pullout capacity of anchor plates was by Mors (1959), which proposed a failure surface in the soil at ultimate load which could be approximated as a truncated cone having an apex angle α equal to $(90° + \varphi/2)$ as shown in Fig. 2.8. The net ultimate pullout capacity was assumed to be equal to the weight of the soil mass bounded by the sides of the cone and the shearing resistance over the failure area surface was ignored.

$$P_u = \gamma V \tag{2.5}$$

where

V = volume of the soil in the truncated

γ = unit weight of soil

This information provides guidance for the design and evaluation of anchor systems used to prevent the sliding and/or overturning of laterally loaded structures founded in soils. The typical system of forces acting on a simple anchor is shown in Fig. 2.9. The pullout force is given by the typical equation:

$$P_u + P_s + W + P_t \tag{2.6}$$

where

P_u = ultimate pullout force

w = effective weight of soil located in the failure zone

P_s = shearing resistance in the failure zone

P_t = force below the area

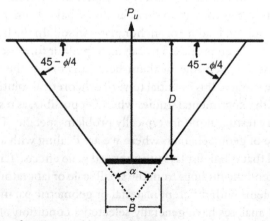

Fig. 2.8 Failure surface assumed by Mors (1959).

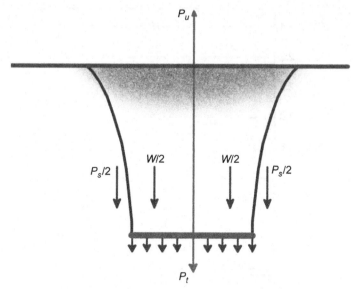

Fig. 2.9 Anchor under pullout test.

In this case involving sands, P_t is equal to zero.

Balla (1961) proposed a method to predict the ultimate pullout capacity of an anchor plate. Balla developed a shearing resistance model during failure surface that involved:

$$P_u = H^3 \gamma \left[F_1 \left(\varphi \frac{L}{D} \right) + F_3 \left(\varphi, \frac{L}{D} \right) \right] \tag{2.7}$$

The sum of F_1, F_3 can be seen in Fig. 2.10. The breakout factor is defined as:

$$N_q = \frac{P_u}{\gamma A H} \tag{2.8}$$

Downs and Chieurzzi (1966), based on similar theoretical work, investigated an apex angle always equal to 60 degrees, irrespective of the friction angle of the soil. But Teng (1962) and Sutherland (1988) found that this assumption might lead to unsafe conditions in many cases common with increase in depth. An approximate semiempirical theory for the pullout loading force of horizontal strip, circular, and rectangular anchors has been proposed by Meyerhof and Adams (1968) (Fig. 2.11).

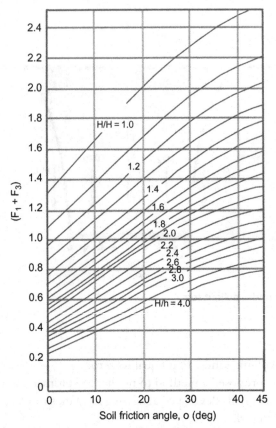

Fig. 2.10 Variation of F1 + F3 based on Balla's result (1961).

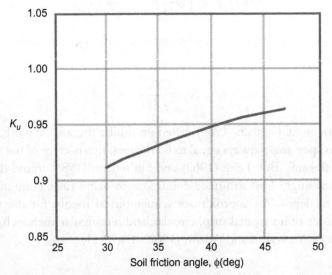

Fig. 2.11 Variation of K_u based on Meyerhof and Adams (1968).

For a strip anchor, an expression for the ultimate pullout capacity was selected by considering the equilibrium of the block of soil directly above the anchor (i.e., contained within the zone made when vertical planes are extended from the anchor edges). The capacity was assumed to act along the vertical planes extending from the anchor shape, while the total passive earth pressure was assumed to act at some angle to these vertical planes. This angle was selected based on laboratory test results while the passive earth pressures were evaluated from the results of Caquot and Kerisel (1949). For shallow plate anchors where the failure surface develops to the soil surface, the ultimate pullout capacity was determined by considering the equilibrium of the material between the anchor and soil surface. For a deep anchor the equilibrium of a block of soil extending a vertical distance H above the anchor was presented, where H was less than the actual embedment depth of the plate anchor. The magnitude of H was determined from the observed extent of the failure surface from laboratory works.

The analysis of strip footings was developed by Meyerhof and Adams to include circular plate anchors by using a semiempirical shape factor to modify the passive earth pressure obtained for the plane strain case. The failure surface was assumed to be a vertical cylindrical surface through the anchor edge and extending to the soil surface. An approximate analysis for the capacity of rectangular plate anchors was selected as for downward loads (Meyerhof 1951), by assuming the ground pressure along the circular perimeter of the two end portions of the failure surface was governed by the same shape factor assumed for circular anchors. It was, however, based on two key adoptions: namely, the edge of the failure surface and the distribution of stress along the failure surface. Even so, the theory presented by Meyerhof and Adams (1968) has been found to give reasonable estimates for a wide range of plate anchor problems. It is one of only two methods available for appraising the force of rectangular plate anchors (Fig. 2.12).

Meyerhof and Adams (1968) expressed the ultimate pullout capacity in rectangular anchor plates as the following equation:

$$P_u = W + \gamma H^2 \left(2 S_f L + B - L\right) K_u \tan \varphi \qquad (2.9)$$

$$S_f = 1 + m \frac{L}{D} \qquad (2.10)$$

$$N_q = 1 + \frac{L}{D} K_u \tan \varphi \qquad (2.11)$$

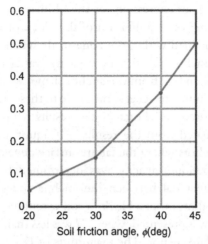

Fig. 2.12 Variation of *m* based on Meyerhof and Adams (1968).

Vesic (1971) studied the problem of an explosive point charge expanding a spherical close to the surface of a semiinfinite, homogeneous and isotropic soil (Figs. 2.13 and 2.14).

$$P_u = \gamma H A N_q \tag{2.12}$$

$$N_q = \left[1 + A_1 \left(\frac{H}{\frac{h_1}{2}} \right) + A_2 \left(\frac{H}{\frac{h_1}{2}} \right)^2 \right] \tag{2.13}$$

Clemence and Veesaert (1977) showed a formulation for shallow circular anchors in sand assuming a linear failure making an angle of $\beta = \varphi/2$ with the

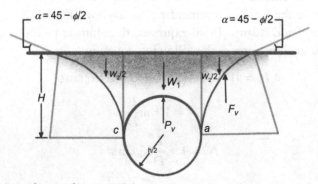

Fig. 2.13 View of tests of Vesic (1971).

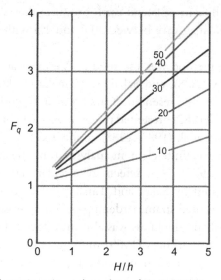

Fig. 2.14 Breakout factor in strip anchor plate of Vesic (1971).

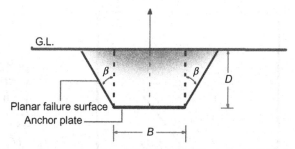

Fig. 2.15 Failure surface assumed by Clemence and Veesaert (1977).

vertical through the shape of the anchor plate as shown in Fig. 2.15. The contribution of shearing resistance along the length of the failure surface was approximately taken into consideration by selecting a suitable value of ground pressure coefficient from laboratory model works. The net ultimate pullout capacity can be given as

$$P_u = \gamma V + \pi \gamma K_o \tan\varphi \cos^2\left(\frac{\phi}{2}\right)\left(\frac{BD^2}{2} + \frac{D^3 \tan\frac{\phi}{2}}{3}\right) \qquad (2.14)$$

where:

V was the volume of the truncated cone above the anchor, and

K_o was the coefficient of lateral earth pressure; they suggested that the magnitude of K_o may vary between 0.6 and 1.5 with an average value of about 1.

The finite element method was also used by Vermeer and Sutjiadi (1985), Tagaya et al. (1983, 1988), and Sakai and Tanaka (1998). Unfortunately, only limited results were presented in these research works.

Rowe and Davis (1982) presented research on the behavior of an anchor plate in sand. Tagaya et al. (1983, 1988) conducted two-dimensional plane strain and axisymmetric finite element analyses using the constitutive law of Lade and Duncan (1975). Scale effects for circular plate anchors in dense sand were investigated by Sakai and Tanaka (1998) using a constitutive model for a nonassociated strain hardening–softening elastoplastic material. The effect of shear band thickness was also introduced (Fig. 2.16).

Koutsabeloulis and Griffiths (1989) investigated the trapdoor problem using the initial stress finite element method. Both plane strain and axisymmetric research were conducted.

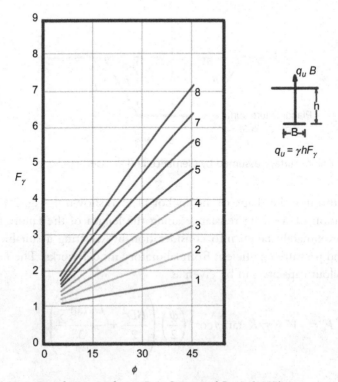

Fig. 2.16 Variation of capacity factor $F\gamma$ in Rowe and Davis (1982).

The researchers concluded that an associated flow rule has little effect on the collapse load for strip plate anchors but a significant effect (30%) for circular anchors. Large displacements were observed for circular plate anchors prior to collapse. In the limit equilibrium method (LEM), an arbitrary failure surface is adopted along with a distribution of stress along the selected surface. Equilibrium conditions are then considered for the failing soil mass and an estimate of the collapse load is assumed. In the research of horizontal anchor force, the failure mechanism is generally assumed to be log spiral in edge (Saeedy, 1987; Sarac, 1989; Murray and Geddes, 1987; Ghaly and Hanna, 1994b) and the distribution of stress is obtained by using either Kotter's equation (Balla, 1961), or by using an assumption regarding the orientation of the resultant force acting on the failure plane. The function of Murray and Geddes (1987) involves:

$$N_q = 1 + \frac{L}{D}\tan\varphi\left(1 + \frac{D}{B} + \frac{\pi L}{3B}\right) \qquad (2.15)$$

Upper and lower bound limit analysis techniques have been studied by Murray and Geddes (1987), Basudhar and Singh (1994) and Smith (1998) to estimate the capacity of horizontal and vertical strip plate anchors. Basudhar and Singh (1994) selected estimates using a generalized lower-bound procedure based on finite elements and nonlinear programming similar to that of Sloan (1988). The solutions of Murray and Geddes (1987) were selected by manually constructing cinematically admissible failure mechanisms (upper bound), while Smith (1998) showed a novel rigorous limiting stress field (lower bound) solution for the trapdoor problem.

2.3.2 Recent Numerical Work

Only a few investigations into the performance of ultimate pullout loading in sand were recorded in model numerical studies. An example of this is work by Fargic and Marovic (2003) that discussed the pullout capacity of plate anchors in soil under applied vertical force. Computation of the pullout and uplift force was performed by the finite element method (FEM). For a gravity load, the concept of initial stresses in Gauss points was selected. In the first increment of computation, these stresses were added to the vector of total stress. The soil was modeled by an elastoplastic constituent material model and the associated flow rule was used. The soil mechanics parameters of samples were determined by standard tests conducted on disturbed samples. For a complex constitutive numerical model of material to describe an actual state of soil, a greater number of soil mechanics parameters must be available. The tensile strength of the soil materials was crucial only in few cases, and the problem of tensile plate anchors is one of them. An iterative procedure was used as the first procedure. The elements with tensile stresses were excluded from the following steps by diminishing the different modulus. More sophisticated constitutive laws are required for an exact analysis, and an adequate finite element method code program has to be prepared.

Merifield and Sloan (2006) used many numerical solutions for analysis of plate anchors. Until this time very few rigorous numerical analyses had been performed to determine the pullout capacity of plate anchors in sand. Although it is essential to verify theoretical solutions/numerical analysis with experimental studies wherever possible, results selected from their laboratory testing alone were typically problem specific. It was particularly the case in geotechnical studies, where they were dealing with a highly nonlinear material that sometimes displays pronounced scale effects (Figs. 2.17–2.19).

As the cost of performing laboratory work on each and every field problem combination is prohibitive, it is necessary to be able to model soil pullout loading numerically for the purposes of design. Existing numerical analyses generally assume a condition of plane strain for the case of a continuous strip plate anchor or axisymmetry for the case of circular plate anchors. The researchers were unaware of any three-dimensional numerical analyses to ascertain the effect of plate anchor shape on the uplift capacity.

Dickin and Laman (2007) investigated the numerical modeling of plate anchors using PLAXIS, which is finite element software. The numerical models were basically 0.57 m long, 0.46 m wide, and 0.23 m deep. Numerical analysis research investigated the uplift response of 1-m wide strip anchors in sand; the results indicated maximum ultimate pullout capacity increase with anchor embedment ratio and sand packing. The research was carried out using a plane strain model for anchors in both loose and

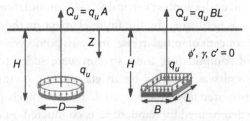

Fig. 2.17 Problem definition by Merifield and Sloan (2006).

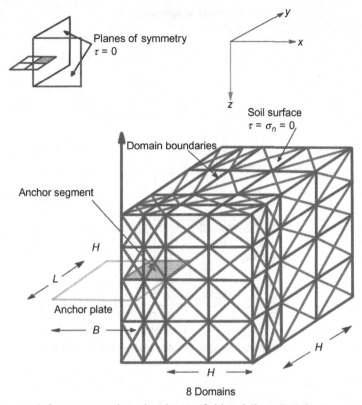

Fig. 2.18 Mesh for square anchor plate by Merifield and Sloan (2006).

dense sand. During the generation of the mesh, 15–node triangular elements were obtained in the determination of stresses (Figs. 2.20–2.23).

Kumar and Kouzer (2008) used a group of multiple strip plate anchors placed in sand and subjected to equal magnitudes of vertical upward pullout loads determined by numerical solutions. Instead of using a number of anchor plates in numerical modeling, a single plate anchor was used by modeling the boundary conditions along the plane of symmetry on both sides of the plate anchor. The effect of interference due to a number of multiple strip plate anchors placed in a granular medium at different embedment depths was investigated by conducting a series of small numerical models (Fig. 2.24).

Kuzer and Kumar (2009) used a group of two spaced strip plate anchors to study the vertical pullout loading of two interfering rigid rough strip

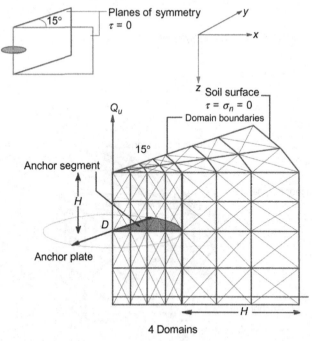

Fig. 2.19 Mesh for circular anchor plate by Merifield and Sloan (2006).

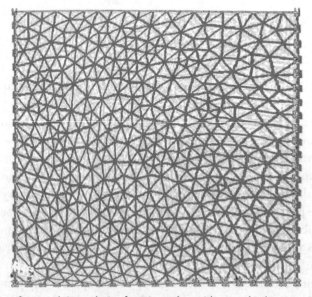

Fig. 2.20 Very fine mesh in analysis of a strip anchor with an embedment ratio $L/D=7$ in PLAXIS by Dickin and Laman (2007).

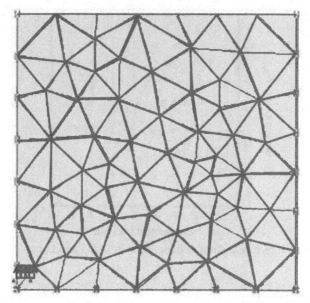

Fig. 2.21 Coarse mesh in analysis of a strip anchor with an embedment ratio $L/D = 7$ in PLAXIS by Dickin and Laman (2007).

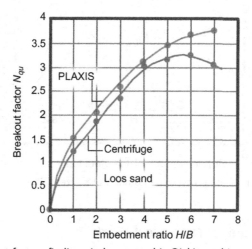

Fig. 2.22 Breakout factors findings in loose sand in Dickin and Laman (2007).

anchors embedded horizontally in sand. The analysis was performed by obtaining an upper-bound theorem of limit analysis combined with finite element and linear programming. The authors used an upper-bound finite element limit analysis; the efficiency factor ξ_γ was computed for a group of two closely spaced strip plate anchors in sand (Fig. 2.25).

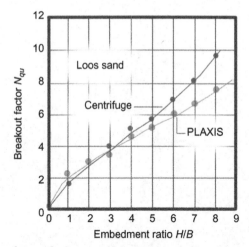

Fig. 2.23 Breakout factors findings in dense sand in Dickin and Laman (2007).

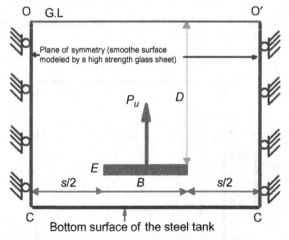

Fig. 2.24 Definition of the problem by Kumar and Kouzer (2008).

2.3.3 Limitations of Numerical Work

Some of the previous researchers reported dealing with analysis of the limiting different numerical methods in ultimate pullout capacity. Very few rigorous numerical studies have been undertaken to determine anchor behavior. It is generally agreed that existing theories do not describe the behavior of anchor plates well enough (Sutherland (1988)). Most methods of analysis are based upon the initial assumption of a common failure mode

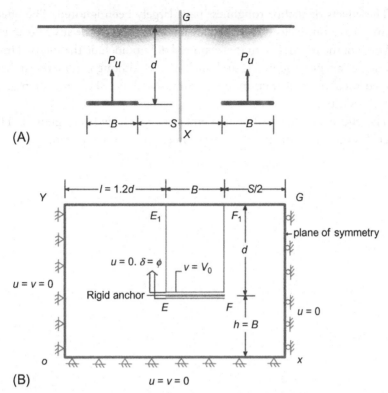

Fig. 2.25 (A) Position and loading of anchors; (B) boundary condition by Kuzer and Kumar (2009).

(limit equilibrium method and upper-bound limits analysis). Given that few attempts have been made to accurately monitor temporary soil deformations under laboratory conditions, the validity of the assumed failure mechanisms remain largely unproven. Different plate anchors and soil parameters have been employed by different researchers. Inevitably such a wide range of parameters will contribute to conflicting conclusions for ultimate pullout loading of anchor plates.

Most anchor studies have been concerned with either vertical or horizontal uplift resistance. However, anchors and researchers are frequently placed at orientations somewhere between horizontal and vertical, depending on the type of application and load orientation (i.e., seawall, retaining walls and transmission tower foundations). The effect of plate anchor inclination on the pullout capacity needs to be investigated.

The effects of anchor roughness have largely been ignored. The effect this may have on shallow vertical plate anchors needs to be investigated.

Some of the research activities reported did not include the internal friction angle, anchor roughness and anchor size, although most researchers selected their internal friction angle using the direct shear test or triaxial compression test.

The effects of plate anchor movement have largely been ignored. The effect this may have on soil deflections needs to be investigated.

CHAPTER 3

Research Methodology

3.1 INTRODUCTION

Previous researchers adopted full-scale, small-scale or centrifuge modeling techniques to determine the ultimate pullout capacity of a plate anchor in sand. Researchers such as Ovesen (1981), Dickin (1988), Tagaya et al. (1988), and Dickin (1994) employed a centrifugal technique to replicate a prototype plate anchor in establishing the ultimate pullout capacity. The works of Hanna and Carr (1971), Das and Seeley (1975a,b), Sakai and Tanaka (1998), and Pearce (2000) were mainly based on small model tests in the laboratory. In the present research, model testing was carried out on small and large irregular shape anchors performed in a chamber box.

3.2 TESTING PROGRAM

The tests performed on small and large irregular shape anchors, with lengths of 297 and 159 mm, were conducted in a chamber box with loose and dense sand packing conditions. The irregular shaped anchor models were employed with an embedment ratio L/D between 1–7 for the small model and 1–4 for the large model. For a basis of comparison, results were compared to previous theoretical results.

3.3 CHAMBER BOX ARRANGEMENT

The chamber box is a large container measuring 1400 mm in length, 700 mm in width and 1500 mm in depth. It was fabricated using reinforced steel angle 150 mm × 150 mm × 60 mm as framed and closed by plywood of 25 mm in thickness, which was then laminated by a Formica layer to reduce the friction effect. For leveling purposes and reducing the influence of humidity from the floor, the chamber was supported by C-channel of 200 mm in size. The chamber box is shown in Fig. 3.1. A motor displacement speed of 60 mm/min and a load cell were employed in the chamber box; they are shown in Figs. 3.2–3.4.

Irregular Shape Anchor in Cohesionless Soils
http://dx.doi.org/10.1016/B978-0-12-809550-8.00003-4

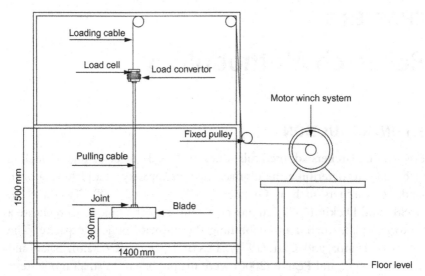

Fig. 3.1 View of pullout test.

Fig. 3.2 View elevation of data logger.

3.4 TESTING PROCEDURE IN CHAMBER BOX

The sand was in both dense and loose packing. Initially a sand layer of 300 mm thickness was placed in a chamber box as loose sand. The loose packing procedure was accomplished by raining sand from the top of the chamber box proposed for loose packing. A unit weight of 14.90 kN/m^3

Fig. 3.3 View of motor winch system.

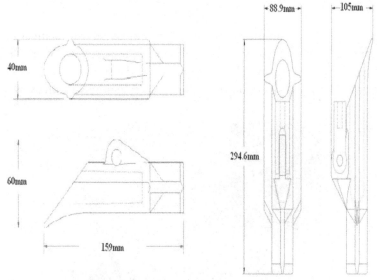

Fig. 3.4 View of irregular shape anchor details.

was achieved for this type of sand packing. A rectangular container, with a hole arranged in a rectangular grid form, was placed at a height of 1200 mm on top of the chamber box for the purpose of loose packing. For dense packing, the sand was compacted using an electrical vibrator at every 80-mm thickness for 8 min, covering all surfaces.

3.5 EXPERIMENTAL PROGRAM

3.5.1 Properties of Sand

Dry and clean sand was dried in the ovens with a temperature of 100°C for about 24 h to ensure its dry condition. The grain size sand is between 0.07 and 4.99 mm, with a mean grain size $D_{50} = 0.65$. Figs. 3.5 and 3.6 show the grain size distribution of the sand used in the experimental test. The coefficient of uniformity of sand C_u is approximately 3.76, with specific gravity $G_s = 2.71$. The sand is classified as well graded with effective grain size $D_{10} = 0.17$ mm.

The shear strength parameters of the sand were obtained from a direct shear test. Figs. 3.7 and 3.8 show the result of a shear direct test. The average results are 42° and 35° for dense and loose packing sand, respectively.

Relative porosity gives a better assessment on packing condition than the porosity alone. Kolbuszewski (1946) defined the relative porosity as:

$$n_r = \frac{n_{max} - n}{n_{max} - n_{min}} \tag{3.1}$$

where n_{max} and n_{min} are the limiting porosities obtained from standard procedures to achieve very loose and dense packing, respectively, while n is the actual porosity. Apart from relative porosity, the density index I_d is also used to describe the packing condition of the sand. It can be written as:

Fig. 3.5 View of sieve test.

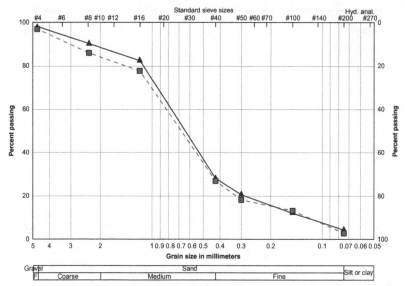

Fig. 3.6 Variations of grain size distribution of sand.

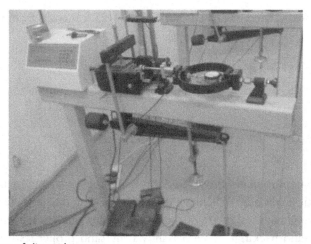

Fig. 3.7 View of direct shear test.

$$I_d = \frac{e_{\max} - e}{e_{\max} - e_{\min}} \tag{3.2}$$

where $e_{\max}$ and $e_{\min}$ are the limiting void ratios, which refer to loose and dense conditions, respectively, while e is the actual void ratio.

The procedure for obtaining maximum porosity $n_{\max}$ can be conducted by carrying out the maximum void ratio $e_{\max}$ and, by using soil mechanics principles, the values of porosity and density index were found. The limiting

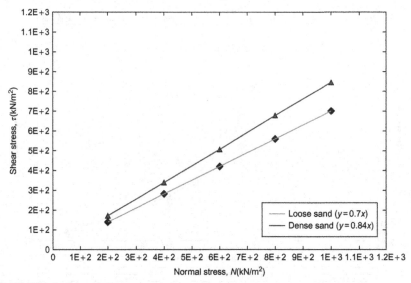

Fig. 3.8 Variations of shear stress with normal stress.

void ratio was carried out based on ASTM D4253. Results from these tests include:

$$e_{max} = 0.832$$
$$e_{min} = 0.594$$
$$n_{max} = 0.462$$
$$n_{min} = 0.382$$
$$\gamma_{d\,max} = 14.90 \text{kN/m}^3$$
$$\gamma_{d\,min} = 16.95 \text{kN/m}^3$$

A sand-raining method was used to obtain a loose packing while hand-held vibration was used to achieve the dense packing. From these methods, dry unit weights of soil of 16.95 and 14.90 kN/m³ were achieved in dense and loose sand packing. Table 3.1 shows the sand properties used in this research.

3.5.2 Irregular Shape Anchor Materials

Two different sizes of irregular shape anchor were designed in mechanical software (ANSYS – CATIA v.5). The design steps and results of the irregular shape anchor are as shown in Appendix A. Small and large irregular shape anchors were fabricated from galvanized steel with lengths of 297 and

Table 3.1 Parameters of sample tested

Parameter	Index	Value
Specific gravity	G_s	2.71
Maximum void ratio	e_{max}	0.832
Minimum void ratio	e_{min}	0.594
Coefficient of uniformity	$C_u = D_{60}/D_{10}$	3.76
Effective size	D_{10}	0.17 mm
Different size	D_{50}	0.65 mm
Maximum porosity	n_{max}	0.462
Minimum porosity	n_{min}	0.382
Dry unit weight of loose sand	γ_{dmin}	14.90 kN/m^3
Dry unit weight of dense sand	γ_{dmax}	16.95 kN/m^3

159 mm, widths of 88 and 63 mm and special shape and format. The small size irregular shape anchor had a length of 159 mm and width 63 mm and did not have a wing area from galvanized steel. The surfaces of the irregular shape anchor were maintained as rough.

A cable is used as a pulling rod for pullout tests of the irregular shape anchor in the chamber box. The pulling rod was extended above the top of the irregular shape anchor and a load cell was attached in a special steel frame by using pullout loading.

3.5.3 Chamber Box

The irregular shape anchors were tested in the chamber box and their results were compared to previous theoretical results. The irregular shape anchors have lengths of 159 and 297 mm for the small and large models. The models had an embedment ratio between 1–7 in the small model and 1–4 in the large model.

A large irregular shape anchor and load cell was employed in the chamber box, shown in Fig. 3.9.

3.5.4 Failure Mechanism

The failure mechanism tests were performed as shown in Fig. 3.10. In these tests, patterns were made on the extreme pullout loads and embedment ratio. The aim of these tests was to show the behavior of the failure mechanism of loose and dense sand around an irregular shape anchor due to pullout loading.

The properties of the test were applied to a unit of weight of 16.95 kN/m^3 and 14.90 sand^3 obtaining dense sand and loose sand. Every

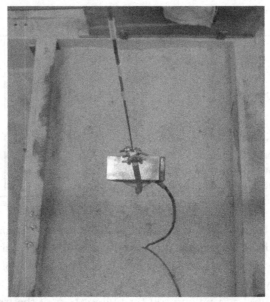

Fig. 3.9 View of testing still running.

Fig. 3.10 View of failure mechanism.

40 mm, vertical intervals of 4 mm were involved, dyed in such a way that sand was placed on the front face of the failure box in alignment with visual line. Loading was applied to the small and large irregular shape anchor through the loading cable with a constant rate of loading. The failure pattern was shown during the testing. The irregular shape anchors

were made to move until sufficient distance was achieved, to ensure the failure pattern was shown. The results of these tests are given in Chapter 4.

3.6 BREAKOUT FACTOR

The main parameters of the collapse load, which may act on geotechnical parameters, are those due to the unit weight of sand, internal friction, irregular shape anchor embedded depth and size of the irregular shape anchor. In full-scale model analysis, the equation of those parameters may be expressed in dimensionless form as follows:

$$f_1(P, L, D, \varnothing, \gamma) = 0 \tag{3.3}$$

f_1 may be expressed as f_2, where

$$f_2(\pi_1, \pi_2, \pi_3) = 0 \tag{3.4}$$

Since $\varnothing$ is dimensionless, thus

$$\pi_1 = \varnothing \tag{3.5}$$

Then,

$$P = f(L, D, \gamma)$$
$$P = L^\alpha D^\beta \gamma^c$$
$$MLT^{-2} = (L)^\alpha (L)^\beta (ML^{-2}T^{-2})^c$$
$$\alpha = 1, \ \beta = 2, \ c = 1$$

Then,

$$P = LD^2\gamma$$
$$\pi_2 = P/LD^2\gamma \tag{3.6}$$

L and D have the same dimensional form, so

$$\pi_3 = L/D \tag{3.7}$$

Thus,

$$f_1\left(\varnothing, P/LD^2\gamma, L/D\right) = 0 \tag{3.8}$$

$$\frac{P}{LD^2\gamma} = f\left(\varnothing, \frac{L}{D}\right)$$
$$P = f\left(\varnothing, \frac{L}{D}\right) \times LD^2\gamma \tag{3.9}$$

where P is the ultimate pullout load obtained from the test, D is the width of the irregular shape anchor, H is embedded depth of the irregular shape anchor, γ is dry unit weight, $\emptyset$ is the internal friction angle and L/D is the embedment ratio. The internal friction angle is the constraint for the test.

3.7 THEORETICAL ANALYSIS

From the experimental results of this research, an empirical formula was created based on breakout factor and embedment ratios in any conditions. Breakout factors were obtained to ultimate pullout capacity. The relationship between breakout factors and embedment ratios derived from present research and many theoretical approaches for loose sand and dense sand, as shown in Chapter 5. The author's empirical formula was compared to many previous theories such as Balla (1961), Meyerhof and Adams (1968), Vesic (1971), Rowe and Davis (1982), and Dickin and Laman (2007) for the small irregular shape anchor. The author's empirical formula for the large model was also compared to many previous theories, such as Balla (1961), Meyerhof and Adams (1968), Vesic (1971), Rowe and Davis (1982), Murray and Geddes (1987), and Dickin and Laman (2007). The previous theories were the result of the work of many researchers on anchor plates using finite element method analysis, limit equilibrium analysis and limit analysis.

CHAPTER 4

Experimental Results

4.1 INTRODUCTION

Sand was used as an embedment medium in this research. In order to obtain two different criteria for index density, a loose and dense packing were used. Thus, the operation of the irregular shape anchor in both conditions can be observed. The loose packing was achieved by sand-raining methods while the dense packing was achieved through hand compaction via an electrical vibrator. The grain sizes of the sand used in the chamber box have a size ranging from 0.07 to 4.99 mm. Different sized irregular shape anchors were used as well as different embedment ratios (L/D) in the chamber box. Results obtained from the testing conducted in chamber box are discussed in this chapter. The effects of embedment ratio, sand density, shape factor, breakout factor and failure mechanism patterns of models are detailed for the tests. Test data were collected and presented in different curves. The failure mechanism patterns of the models in loose sand and dense sand were observed from the tests and are explained in this chapter.

4.2 IRREGULAR SHAPE ANCHOR IN CHAMBER BOX

Two different sized irregular shape anchors, one 157 mm in length and 40 mm in width and the other 297 mm in length and 88.9 mm in width, were tested in the chamber box.

4.2.1 Irregular Shape Anchor in Chamber Box

Factors involving the determination of ultimate pullout capacity were observed. Embedment ratio L/D, sand density γ, model size D, and breakout factor N_q were all correlated with pullout loads.

4.2.1.1 The Effect of Embedment Ratio

Summaries of test results on the effect of embedment ratio are tabulated as shown in the figures. These results are presented graphically in Figs. 4.1–4.4 illustrating maximum model pullout load against displacement of between 1–7 for the small model and 1–4 for the large model. The pullout load

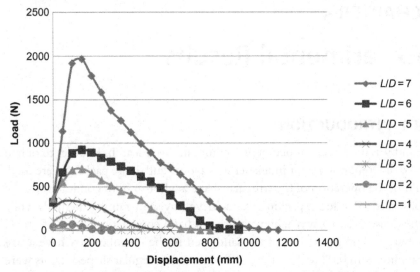

Fig. 4.1 Variations of pullout load with embedment ratio *L/D* for small irregular shape anchor in loose sand.

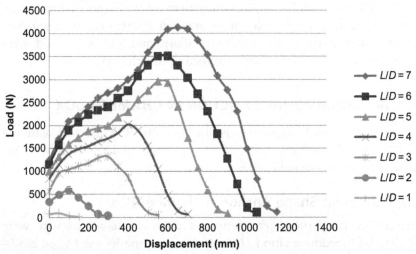

Fig. 4.2 Variations of pullout load with embedment ratio *L/D* for small irregular shape anchor in dense sand.

increases significantly with an increase in embedment ratio. The maximum small model pullout load shows an almost similar trend when the embedment ratio was at $L/D=7$. Figs. 4.1 and 4.3 illustrate the influence of embedment ratio L/D for the 159 mm and 297 mm lengths of the models

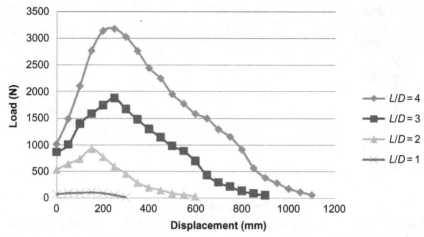

Fig. 4.3 Variations of pullout load with embedment ratio L/D for large irregular shape anchor in loose sand.

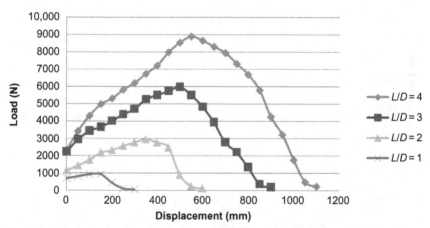

Fig. 4.4 Variations of pullout load with embedment ratio L/D for large irregular shape anchor in dense sand.

in loose sand, respectively. Fig. 4.3 shows that for an embedment ratio $L/D < 4$, the maximum model of pullout load increases until the embedment ratio $L/D = 4$. Figs. 4.2 and 4.4 illustrate the influence of an embedment ratio L/D for 159 mm and 297 mm lengths of models in dense sand, respectively.

Generally, the maximum of the pullout load increases with an increase in embedment ratio L/D. Figs. 4.5 and 4.6 illustrate the influence of the

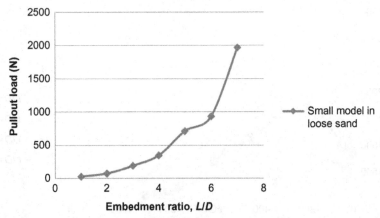

Fig. 4.5 Result of pullout load with embedment ratio *L/D* for small irregular shape anchor in loose sand.

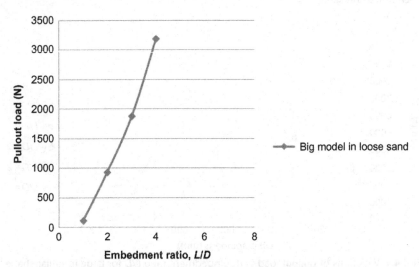

Fig. 4.6 Result of pullout load with embedment ratio *L/D* for large irregular shape anchor in loose sand.

embedment ratio L/D for small and large irregular shape anchors in loose sand, respectively. Fig. 4.5 shows that for an embedment ratio $L/D > 1$, the maximum pullout load increases until the embedment ratio $L/D = 7$ and Fig. 4.6 shows that for an embedment ratio $L/D > 1$, the maximum model of pullout load increases until the embedment ratio $L/D = 4$. This shows that the embedment ratio has a significant effect on the contribution of increasing pullout load in loose sand.

The maximums of the pullout load with an embedment ratio L/D between 1–4 for the large model and 1–7 for the small model in dense sand are illustrated in Figs. 4.7 and 4.8. They show the influence of embedment ratio L/D on small and large irregular shape anchors in dense sand,

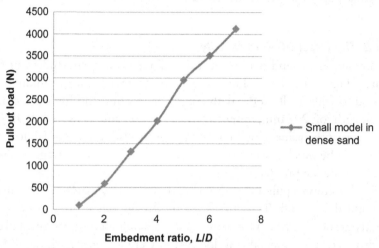

Fig. 4.7 Result of pullout load with embedment ratio L/D for small irregular shape anchor in dense sand.

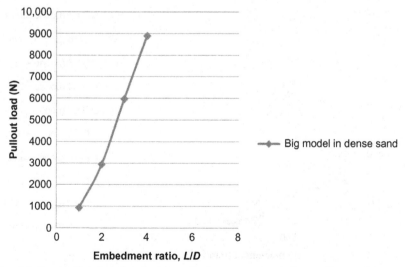

Fig. 4.8 Result of pullout load with embedment ratio L/D for small irregular shape anchor in dense sand.

respectively. Fig. 4.7 shows that for an embedment ratio $L/D > 1$, the maximum pullout load increases until the embedment ratio $L/D = 7$ and Fig. 4.8 shows that for an embedment ratio $L/D > 1$, the maximum model of pullout load increases until the embedment ratio $L/D = 4$ in dense sand. This shows that the embedment ratio has a significant effect on the contribution of increasing pullout load in dense sand.

4.2.1.2 The Effect of Sand Density

The influences of sand density on the pullout loading are studied in this research. Figs. 4.9 and 4.10 show a comparison between the maximum pullout load for the small model with length of 159 mm and the large model with length of 297 mm, respectively, in loose and dense sand. It was expected that the value of pullout loading for small and large models in loose sand would be much lower compared with those of similar irregular shape anchors embedded in dense sand.

Fig. 4.9 shows a pullout for the small model with a length of 159 mm in loose and dense sand. The curves for model pullout in loose sand were expected to be lower than for dense sand. The pullout loads for small irregular shape anchors embedded in loose sand were lower than those of their equivalent in dense sand.

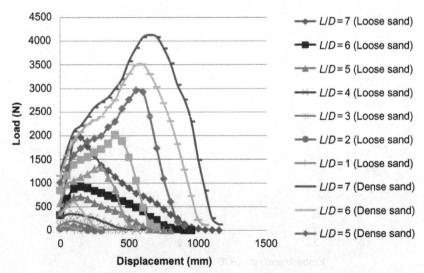

Fig. 4.9 Variations of pullout load with embedment ratio L/D for small irregular shape anchor in loose and dense sand.

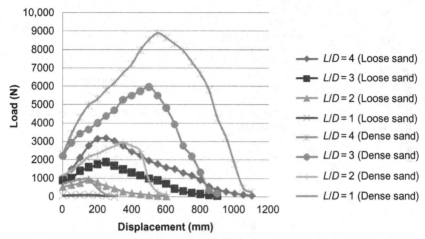

Fig. 4.10 Variations of pullout load with embedment ratio *L/D* for large irregular shape anchor in loose and dense sand.

Fig. 4.10 shows experimental data for the large model with a length of 297 mm in loose and dense sand. The resulting curves in loose sand were expected to be lower than in dense sand. The pullout loads for the large irregular shape anchor embedded in loose sand were lower than those of their equivalent in dense sand.

4.2.1.3 Breakout Factor

The influences of breakout are studied in this research. Fig. 4.11 shows a comparison between breakout factor for the small model with a length of 159 mm and the large model with a length of 297 mm, respectively, in loose sand. It was expected that the value of breakout factor would increase with the embedment ratio as shown in Figs. 4.11 and 4.12. It was expected that the value of breakout factor for small and large models in loose sand would be much lower compared with those of similar irregular shape anchors embedded in dense sand.

Fig. 4.12 shows a model of breakout factor for small and large models in dense sand. The result of breakout in dense sand was expected to be more than in loose sand.

4.3 FAILURE MECHANISM

Tests were carried out in the glass box to examine the failure mechanism for irregular shape anchor embedment in loose and dense sand. The small

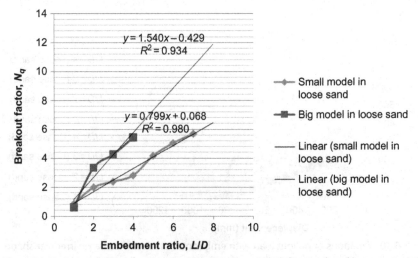

Fig. 4.11 Variations of breakout with embedment ratio L/D for small and large irregular shape anchor in loose sand.

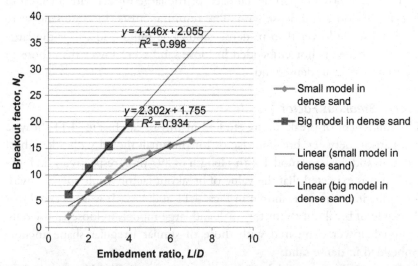

Fig. 4.12 Variations of breakout with embedment ratio L/D for small and large irregular shape anchor in dense sand.

irregular shape anchor was selected for this test. The sand was used to form multiple specific layers of lines to observe the movement of the loose sand and dense sand due to pullout of the small irregular shape anchor. The shape line deformation at the small model was recorded by camera in real time. The colored tapes show the balanced surface on the side of the small irregular shape anchor before being subjected to loading. The method of loading sand was similar to that for the main tests. The failure mechanism is shown, conducted using irregular shape anchors and their rotation with a specified embedment ratio L/D. The results of the failure mechanism due to pullout loading of the small model in loose sand conditions are shown in Figs. 4.13–4.20. Figs. 4.13–4.18 show the result of pullout loading of the small model in dense sand.

4.3.1 Failure Mechanism in Loose Sand

Figs. 4.13–4.20 show the failure mechanism of the small irregular shape anchor in loose sand. Figs. 4.17–4.20 illustrate the small irregular shape

Fig. 4.13 Failure mechanism setup.

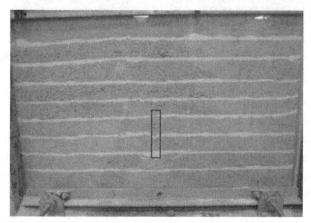

Fig. 4.14 Front elevation for failure mechanism test in loose sand.

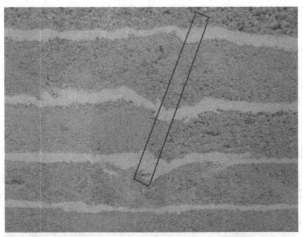

Fig. 4.15 Front elevations for failure mechanism testing in loose sand. Rotation zone starting to develop shear plane, indicated by *lines*.

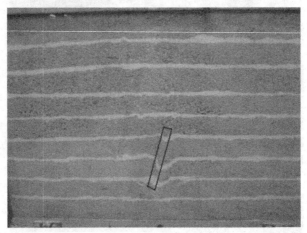

Fig. 4.16 Front elevations for failure mechanism testing in loose sand. Rotation zone indicated by *lines*.

anchor which was subjected to the pullout loading test. The initial movement of the small model shows that the zone was rotated. The sand level in back of the small model starts to rise due to the rotation of the surrounding loose sand and in front of the small model it starts to reduce due to shear of the small model in loose sand. When the pullout loads exceed the surcharge load, the small irregular shape anchor begins to rotate from its initial position. After rotation, the movement of the small model develops a rupture plane within the vicinity of the small irregular shape anchor. The rupture

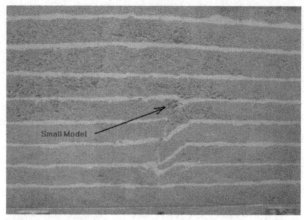

Fig. 4.17 Front elevations for failure mechanism testing in loose sand. Different steps of rotation in small model indicated by *lines*.

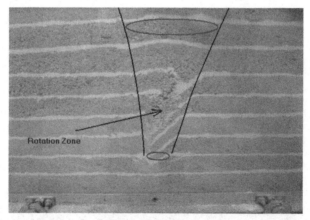

Fig. 4.18 Front elevations for failure mechanism testing in loose sand. Last step of rotation in small model indicated by *lines*.

plane can be clearly seen in Figs. 4.19 and 4.20, showing that it was in agreement with the result of Balla (1961). The pullout load is increased until the ultimate level in the rupture surface extends to the ground surface. Fig. 4.20 shows a full failure mechanism pattern of the small irregular shape anchor subjected to pullout load.

4.3.2 Failure Mechanism in Dense Sand

Figs. 4.21–4.26 show the failure mechanism of the small irregular shape anchor in dense sand. Testing procedures were similar to those for the small

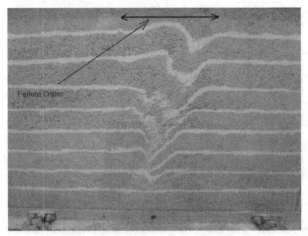

Fig. 4.19 Schematic diagram of failure mechanism in loose sand.

Fig. 4.20 Final schematic of failure zones of small irregular shape anchor in loose sand.

model in loose sand. Figs. 4.24–4.26 illustrate the small irregular shape anchor which was subjected to a pullout loading test. Initial movement of the small model shows that the zone was rotated. The sand level in back of the small model starts to rise due to the rotation of the surrounding dense sand and in front of the small model it starts to reduce, due to shear of the small model in dense sand. The images show the change from dense sand to conditions of loose sand surrounding the small irregular shape anchor during tests. When the pullout loads exceed the surcharge load, the small irregular

Fig. 4.21 View of setup failure mechanism in dense sand.

Fig. 4.22 View of different dense sand layers.

shape anchor begins to rotate from its initial position. After rotation, the movement of the small model develops a rupture plane within the vicinity of the small irregular shape anchor. The rupture plane can be clearly seen in Figs. 4.25 and 4.26, showing that it was similar to the results obtained by Balla (1961). The pullout load is increased until the ultimate level in the rupture surface extends to the ground surface. Fig. 4.26 shows a full failure mechanism pattern of the small irregular shape anchor subjected to pullout load.

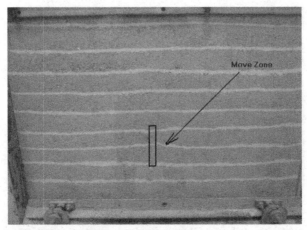

Fig. 4.23 Front elevation for failure mechanism testing of small irregular shape anchor in dense sand, model starting to move indicated by *lines*.

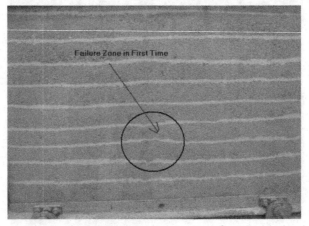

Fig. 4.24 Front elevation for failure mechanism testing of small irregular shape anchor in dense sand, failure zone in first time indicated by *lines*.

4.3.3 Summary of Failure Mechanism

Generally the form of the failure pattern is complex. The shape form of the failure zone in loose sand is smaller than the failure zone in dense sand. The results show that the length of the rupture plane in loose sand is shorter than in dense sand. For the same reasons, the ultimate pullout capacity of the small irregular shape anchor in loose sand was lower than the ultimate pullout capacity in dense sand. The lines of deformation of the small irregular shape

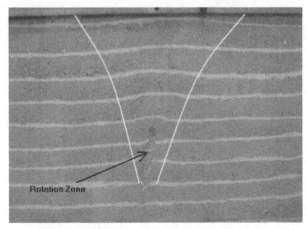

Fig. 4.25 Front elevations for failure mechanism testing in dense sand. Failure zone in small model indicated by *lines*.

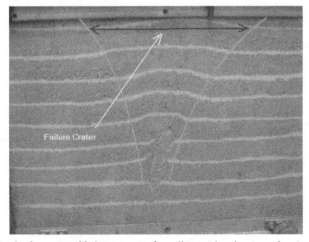

Fig. 4.26 Final schematic of failure zones of small irregular shape anchor in dense sand.

anchor in dense sand were clearer than the lines of deformation of the small irregular shape anchor in loose sand. The results show a similarity of failure pattern of the small irregular shape anchor in both conditions consistent with the findings of Balla (1961). The results show dependence on a failure pattern related to sand conditions involving the friction angle and embedment ratio. The small model created a loose zone into dense sand due to the rotation of model.

CHAPTER 5

Comparison Between Existing Theories and Experimental Works

5.1 INTRODUCTION

Chapter 2 explained previous theoretical research results, which were dedicated to the limiting ultimate pullout capacity of anchor plates, their breakout factor, and failure zones. Researchers such as Balla (1961), Meyerhof and Adams (1968), Sakai and Tanaka (1998), Vesic (1971), Rowe and Davis (1982), Murray and Geddes (1987), Sarac (1989) and Smith (1998), Krishna (2000), Frgic and Marovic (2003), Merifield and Sloan (2006), Dickin and Laman (2007), Kumar and Kouzer (2008), and Kuzer and Kumar (2009) determined the parametric relationships for ultimate pullout capacity of anchor plates and their breakout factor.

Researchers such as Balla (1961), Meyerhof and Adams (1968), Vesic (1971), Rowe and Davis (1982), Murray and Geddes (1987), and Dickin and Laman (2007) dedicated their work to proposing theories for horizontal anchor plates subjected to pullout loads. One of the objectives of their research was to investigate and determine the factors that affect the ultimate pullout capacity of an irregular shape anchor embedded in loose/dense sand.

5.2 ASSUMPTIONS IN COMPARISON WORK

Due to the difficulties in making comparisons, the following assumptions were made:

1. The irregular shape anchors are assumed to have both small and large sizes with an embedment ratio L/D between 1–7 for the small models and 1–4 for the large models.
2. The sand parameters, such as unit weight and internal friction angle, obtained from the experimental activities will be adopted in predicting values from existing theories.
3. Due to limitations of the pullout loading equipment in the laboratory, embedment ratios of 1–4 are used for comparison.

Irregular Shape Anchor in Cohesionless Soils
http://dx.doi.org/10.1016/B978-0-12-809550-8.00005-8

4. The breakout factor will be investigated in this chapter and the results will then be compared to the results of the latest researchers, such as Balla (1961), Meyerhof and Adams (1968), Vesic (1971), Rowe and Davis (1982), Murray and Geddes (1987), and Dickin and Laman (2007).

5.3 COMPARISON OF THE BREAKOUT FACTOR AND EXISTING THEORETICAL BREAKOUT FACTOR

A comparison between the breakout factors of the present research in loose sand and dense sand with breakout factors obtained previously, including Balla (1961), Meyerhof and Adams (1968), Vesic (1971), Rowe and Davis (1982), Murray and Geddes (1987), and Dickin and Laman (2007), is shown in Figs. 5.3 and 5.4. The comparison was constructed based on empirical formulas given by each researcher from their work on anchor plates in loose sand and dense sand.

5.3.1 Comparison of the Breakout Factor and Existing Theoretical Breakout Factor in Loose Sand

The present breakout factor compared with existing breakout factors in loose sand from Balla (1961), Meyerhof and Adams (1968), Vesic (1971), Rowe and Davis (1982), Dickin and Laman (2007) is shown in Fig. 5.1.

According to the experimental results, a close agreement is seen between the author's empirical formula in the small model with that of Balla (1961)

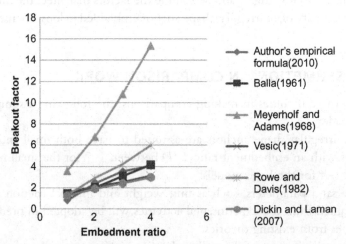

Fig. 5.1 Comparison of present empirical formula in small model with previous theoretical results in loose sand ($\varphi = 35°$).

and Rowe and Davis (1982) for small embedment ratios. Taking into consideration the revisions of Vesic (1971) and Dickin and Laman (2007), the current results are similar to the empirical formulas when the embedment ratio is small. The close difference between the prediction of Meyerhof and Adams (1968) and the empirical formula with embedment ratios having a different value (>1) is due to the rupture surface not extending to the ground surface for anchor plates in a deep position.

The theoretical method that is closest to the physical results is the finite element method of Rowe and Davis (1982). In the design process used by Meyerhof and Adams (1968), a higher safety factor value is required for deep anchor plates, although in the design analysis using Dickin and Laman's (2007) theories, a lower safety factor value is needed.

The comparisons between the present breakout factor and breakout factors in loose sand from Balla (1961), Meyerhof and Adams (1968), Vesic (1971), Rowe and Davis (1982), Murray and Geddes (1987), and Dickin and Laman (2007) are shown in Fig. 5.2.

A close similarity is noted between the author's empirical formula in the small model, based on the physical results, and the Meyerhof and Adams

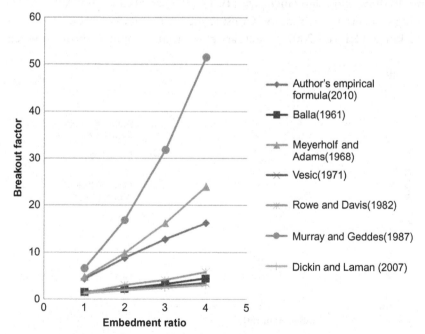

Fig. 5.2 Comparison of present empirical formula in large model with previous theoretical results in loose sand ($\varphi = 35°$).

(1968) results for small embedment ratios. But it should be mentioned that a close agreement also is noted in the case of Murray and Geddes's (1987) predictions with the empirical formula when the embedment ratio had a different value (>2), due to the rupture surface not extending to the ground surface for anchor plates in deep positions.

According to the results, the limited equilibrium method of Meyerhof and Adams (1968) appears to be the best theoretical method, with results close to the physical results. A higher safety factor value is needed for the design by Murray and Geddes (1987) for deep anchor plates, although a lower value is required for the design using the analysis of Dickin and Laman (2007).

5.3.2 Comparison of the Breakout Factor and Existing Theoretical Breakout Factor in Dense Sand

The present breakout factor in dense sand compared with existing breakout factors from Balla (1961), Meyerhof and Adams (1968), Vesic (1971), Rowe and Davis (1982), and Dickin and Laman (2007) is shown in Fig. 5.3.

The author's empirical formula for the small model based on the physical results shows close agreement with Balla (1961) and Rowe and Davis (1982) using small embedment ratios. Considering the revisions of Vesic (1971) and Dickin and Laman (2007), results are closer to the empirical formulas when

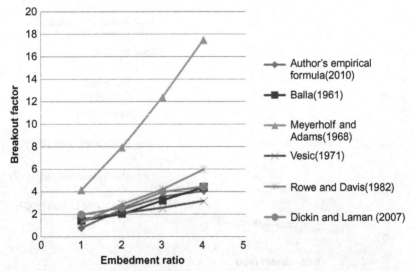

Fig. 5.3 Comparison of present empirical formula in small model with previous theoretical results in dense sand ($\varphi = 42°$).

the embedment ratio was in small. However, the predictions of Meyerhof and Adams (1968) show a significant close difference with the empirical formula with embedment ratios of a different value (>1), due to the rupture surface not extending to the ground surface for anchor plates in the deep position.

According to the results, the finite element method of Rowe and Davis (1982) appears to be the best theoretical method, most closely matching the physical results. A higher value of safety factor is needed for the design using Meyerhof and Adams's (1968) theory for deep anchor plates, although a lower safety factor value is needed for the design using the analysis of Vesic (1971).

A comparison of the present breakout factor with existing breakout factors in dense sand, including Balla (1961), Meyerhof and Adams (1968), Vesic (1971), Rowe and Davis (1982), Murray and Geddes (1987), and Dickin and Laman (2007), is shown in Fig. 5.4.

The author's empirical formula for the small model, based on the physical results, shows close agreement with Meyerhof and Adams (1968) for small embedment ratios. However, with the predictions of Murray and Geddes (1987) is a close difference with the empirical formula when the embedment

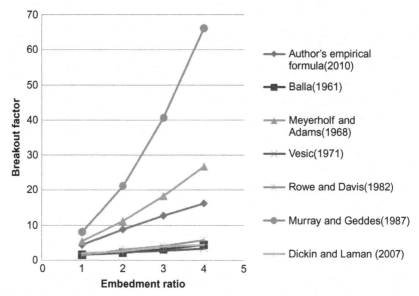

Fig. 5.4 Comparison of present empirical formula in large model with previous theoretical results in dense sand ($\varphi = 42°$).

ratio takes on different values (>2), due to the rupture surface not being extended to the ground surface for anchor plates in the deep position.

According to the results, the limited equilibrium method of Meyerhof and Adams (1968) appears to be the best theoretical method. A higher safety factor value is needed for the design using Murray and Geddes's (1987) theory for deep anchor plates, although a lower safety factor value is needed for the design using the analysis of Vesic's (1971) theory.

CHAPTER 6

Conclusions

6.1 CONCLUSIONS FROM THE RESEARCH

The main purpose of this research was to determine the ultimate pullout capacity and validate the irregular shape anchor based on breakout factor in loose sand and dense sand. The predicted ultimate pullout capacity and the breakout were defined based on empirical formulas from the author's experimental work. Two empirical formulas were developed based on small and large irregular shape anchors in loose sand and dense sand. To account for the ultimate pullout capacity of the small model and the large model, an empirical breakout factor was introduced.

Tests employed in a chamber box showed that the limiting ultimate pullout capacity of irregular shape anchors was influenced by embedment ratio, model sizes and sand density.

The results showed that the patterns of the failure mechanism around the irregular shape anchor at shallow and deep levels were similar to patterns of the rupture surface of Balla (1961) and Dickin and Laman (2007). The results showed that the failure mechanism pattern of loose sand is smaller than that of dense sand.

Empirical formulas were developed by the author based on the experimental results. The lower embedment length gives a lower difference between the empirical formula and previous theories, while an increase in embedment length reduces the deviation. However, Rowe and Davis (1982) appear to become the best solution, being close to the author's experimental results, while Murray and Geddes (1987) with the large model and Meyerhof and Adams (1968) with the small model overpredict the breakout factor for higher embedment ratios. The results showed that breakout factor increased with the size of the irregular shape anchor.

Irregular Shape Anchor in Cohesionless Soils
http://dx.doi.org/10.1016/B978-0-12-809550-8.09985-8

6.2 SUGGESTIONS FOR FUTURE RESEARCH

This keying anchors need to more analysis and investigation in various soils. Irregular Shaped Anchors as a keying anchor need to various analysis in numerical and experimental conditions in different soils. Further suggestions are put forward related to this research for future investigations:

1. Similar research can be done to investigate the ultimate horizontal pull-out load of vertical irregular shape anchors in loose/dense sand.
2. Similar research can be done to investigate the ultimate pullout load of horizontal irregular shape anchors in clay.
3. Similar research can be done to investigate the ultimate pullout load of horizontal irregular shape anchors in different layers.
4. Full-scale tests can be done to research the ultimate pullout load of irregular shape anchors.
5. Numerical analysis can be done during pullout tests around irregular shape anchors by creating new code for the programs.

REFERENCES

Adams, J.I., Hayes, D.C., 1967. The uplift capacity of shallow foundations. Ontario Hydro Res. Q., 1–13.

Akinmusuru, J.O., 1978. Horizontally loaded vertical plate anchors in sand. J. Geotech. Eng. 104 (2), 283–286.

Andreadis, A., Harvey, R., Burley, E., 1981. Embedded anchor response to uplift loading. J. Geotech. Eng. 107 (1), 59–78.

Baker, W.H., Konder, R.L., 1966. Pullout load capacity of a circular earth anchor buried in sand. Highw. Res. Rec. 108, 1–10.

Balla, A., 1961. The resistance of breaking-out of mushroom foundations for pylons. In: Proceedings, 5th International Conference on Soil Mechanics and Foundation Engineering, Paris. vol. 1, pp. 569–576.

Basudhar, P.K., Singh, D.N., 1994. A generalized procedure for predicting optimal lower bound break-out factors of strip anchors. Geotechnique 44 (2), 307–318.

Bouazza, A., Finlay, T.W., 1990. Uplift capacity of plate anchors in a two-layered sand. Geotechnique 40 (2), 293–297.

Caquot, A., Kerisel, L., 1949. Traité de Mécanique des Sols. Gauthier-Villars, Paris.

Clemence, S.P., Veesaert, C.J., 1977. Dynamic pullout resistance of anchors in sand. In: Int. Symp. on Soil Struct Interaction, Roorkee, India, 3 January 1977 through 7 January 1977, pp. 389–397.

Das, B.M., 1990. Earth Anchors. Elsevier, Amsterdam.

Das, B.M., Seeley, G.R., 1975a. Breakout resistance of shallow horizontal anchors. J. Geotech. Eng. ASCE 101 (9), 999–1003.

Das, B.M., Seeley, G.R., 1975b. Load displacement relationship for vertical anchor plates. J. Geotech. Eng. ASCE 101 (7), 711–715.

Dickin, E.A., 1994. Uplift resistance of buried pipelines in sand. Soils Found. 34 (2), 41–48.

Dickin, E.A., 1988. Uplift behaviour of horizontal anchor plates in sand. J. Geotech. Eng. 114 (11), 1300–1317.

Dickin, E.A., Laman, M., 2007. Uplift response of strip anchors in cohesionless soil. J. Adv. Eng. Softw. 1 (38), 618–625.

Dickin, E.A., Leung, C.F., 1983. Centrifuge model tests on vertical anchor plates. J. Geotech. Eng. 109 (12), 1503–1525.

Downs, D.I., Chieurzzi, R., 1966. Transmission tower foundations. J. Power Div. ASCE 88 (2), 91–114.

Frgic, L., Marovic, P., 2003. Pullout capacity of spatial anchors. J. Eng. Comput. 21 (6), 598–700.

Frydman, S., Shamam, I., 1989. Pullout capacity of slab anchors in sand. Can. Geotech. J. 26, 385–400.

Ghaly, A., Hanna, A., 1994a. Model investigation of the performance of single anchors and groups of anchors. Can. Geotech. J. 31 (2), 273–284.

Ghaly, A., Hanna, A., 1994b. Ultimate pullout resistance of single vertical anchors. Can. Geotech. J. 31, 661–672.

Giffels, W.C., Graham, R.E., Mook, J.F., 1960. Concrete cylinder anchors. Electr. World 154, 46–49.

Hanna, T.H., Carr, R.W., 1971. The loading behaviour of plate anchors in normally and over consolidated sands. In: Proceedings, 4th International Conference on Soil Mechanics and Foundation Engineering, Budapest, pp. 589–600.

Hanna, T.H., Sparks, R., Yilmaz, M., 1971. Anchor behaviour in sand. J. Soil Mech. Found. Div. Am. Soc. Civ. Eng. 98 (11), 1187–1208.

Hoshiya, M., Mandal, J.N., 1984. Some studies of anchor plates in sand. Soil. Found., Japan 24 (1), 9–16.

Ireland, H.O., 1963. Uplift resistance of transmission tower foundations: discussion. J. Power Div. ASCE 89 (PO1), 115–118.

Kanakapura, S., Rao, S., Kumar, J., 1994. Vertical uplift capacity of horizontal anchors. J. Geotech. Eng. ASCE 120 (7), 1134–1147.

Kananyan, A.S., 1966. Experimental investigation of the stability of bases of anchor foundations. Osnovanlya, Fundamenty i mekhanik Gruntov 4 (6), 387–392.

Kolbuszewski, J.J., 1946. An experimental study of the maximum and minimum porosities of sands. In: Proc. 2nd Int. Conf. on S.M. and Found Eng. 1, 156.

Koutsabeloulis, N.C., Griffiths, D.V., 1989. Numerical modelling of the trap door problem. Geotechnique 39 (1), 77–89.

Krishna, Y.S.R., 2000. Numerical Analysis of Large Size Horizontal Strip Anchors (Ph.D. Thesis). Indian Institute of Science.

Kumar, J., Kouzer, K.M., 2008. Vertical uplift capacity of horizontal anchors using upper bound limit analysis and finite elements. Can. Geotech. J. 45, 698–704.

Kuzer, K.M., Kumar, J., 2009. Vertical uplift capacity of two interfering horizontal anchors in sand using an upper bound limit analysis. J. Comp. Geotech. 1 (36), 1084–1089.

Lade, P.V., Duncan, J.M., 1975. Elasto-plastic stress-strain theory for cohesionless soil. J. Soil Mech. Found. Div. ASCE 101 (10), 1037–1053.

Mariupolskii, L.G., 1965. The bearing capacity of anchor foundations, SMFE. Osnovanlya, Fundamenty i ekhanik Gruntov 3 (1), 14–18.

Merifield, R., Sloan, S.W., 2006. The ultimate pullout capacity of anchors in frictional soils. Can. Geotech. J. 43 (8), 852–868.

Meyerhof, G.G., 1951. The ultimate bearing capacity of foundations. Geotechnique 2, 301–332.

Meyerhof, G.G., Adams, J.I., 1968. The ultimate uplift capacity of foundations. Can. Geotech. J. 5 (4), 225–244.

Mors, H., 1959. The behaviour of most foundations subjected to tensile forces. Bautechnik 36 (10), 367–378.

Murray, E.J., Geddes, J.D., 1987. Uplift of anchor plates in sand. J. Geotech. Eng. ASCE 113 (3), 202–215.

Niroumand, H., 2007a. Analysis and Modeling with PLAXIS V.8. vols. 1–5. Naghoos Publisher, Tehran. V1:75–128.

Niroumand, H., 2007b. Material Modeling with PLAXIS V.8. vols. 1–5. Naghoos Publisher, Tehran. V3:102–248.

Ovesen, N.K., 1981. Centrifuge tests on the uplift capacity of anchors. Proceedings, 10th International Conference on Soil Mechanics and Foundation Engineering 10 (1), 717–722.

Pearce, A., 2000. Experimental Investigation into the Pullout Capacity of Plate Anchors in Sand (M.Sc Thesis). University of Newcastle, Australia.

Ramesh Babu, R., 1998. Uplift Capacity and Behaviour of Shallow Horizontal Anchors in Soil (Ph.D. Thesis). Dept. of Civil Eng., Indian Institute of Science, Bangalore.

Rowe, R.K., 1978. Soil Structure Interaction Analysis and Its Application to the Prediction of Anchor Behaviour (Ph.D. Thesis). University of Sydney, Australia.

Rowe, R.K., Davis, E.H., 1982. The behaviour of anchor plates in sand. Geotechnique 32 (1), 25–41.

Saeedy, H.S., 1987. Stability of circular vertical anchors. Can. Geotech. J. 24, 452–456.

Sakai, T., Tanaka, T., 1998. Scale effect of a shallow circular anchor in dense sand. Soil. Found., Japan 38 (2), 93–99.

Sarac, D.Z., 1989. Uplift capacity of shallow buried anchor slabs. Proc., 12th Int. Conf. Soil Mech. Found. Eng. 12 (2), 1213–1218.

Sloan, S.W., 1988. Lower bound limit analysis using finite elements and linear programming. Int. J. Numer. Anal. Methods Geomech. 12, 61–67.

Smith, C.C., 1998. Limit loads for an anchor/trapdoor embedded in an associated coulomb soil. Int. J. Numer. Anal. Methods Geomech. 22 (11), 855–865.

Stewart, W., 1985. Uplift capacity of circular plate anchors in sand. Can. Geotech. J. 22 (4), 589–592.

Sutherland, H.B., 1988. Uplift resistance of soils. Geotechnique 38, 493–516.

Sutherland, H.B., 1965. Model studies for shaft raising through cohesionless soils. In: Proceedings, 6th International Conference on Soil Mechanics and Foundation Engineering. vol. 2, pp. 410–413.

Tagaya, K., Scott, R.F., Aboshi, H., 1988. Pullout resistance of buried anchor in sand. Soil. Found., Japan 28 (3), 114–130.

Tagaya, K., Tanaka, A., Aboshi, H., 1983. Application of finite element method to pullout resistance of buried anchor. Soils Found. Japan 23 (3), 91–104.

Teng, W.C., 1962. Foundation Design. Prentice-Hall, Englewood Cliffs, NJ, USA.

Turner, E.Z., 1962. Uplift resistance of transmission tower footings. J. Power Div. ASCE 88 (PO2), 17–33.

Vermeer, P.A., Sutjiadi, W., 1985. The uplift resistance of shallow embedded anchors. In: Proceedings, 11th International Conference on Soil Mechanics and Foundation Engineering, San Francisco, 4, 1635–1638.

Vesic, A.S., 1971. Breakout resistance of objects embedded in ocean bottom. J. Soil Mech. Found. Div. Am. Soc. Civ. Eng. 97 (9), 1183–1205.

Vesic, A.S., 1972. Expansion of cavities in infinite soil mass. J. Soil Mech. Found. Div. Am. Soc. Civ. Eng. 98 (3), 265–290.

Appendices

Appendices

APPENDIX A

Properties of Shear Direct Test in Dense Sand

SETTLEMENT VS SQUARE ROOT TIME

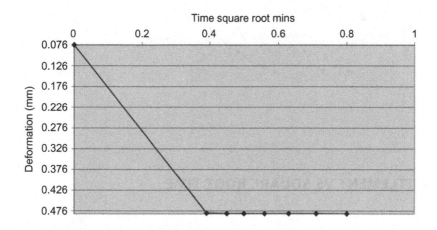

SHEAR STRESS VS DISPLACEMENT

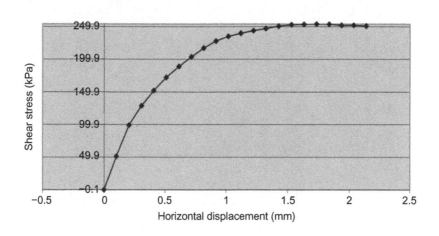

CHANGE IN SPECIMEN THICKNESS VS DISPLACEMENT

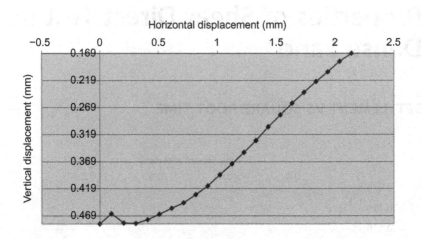

SETTLEMENT VS SQUARE ROOT TIME

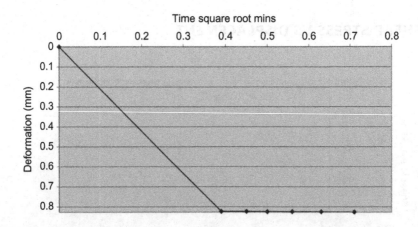

SHEAR STRESS VS DISPLACEMENT

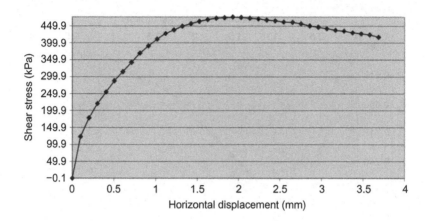

SETTLEMENT VS SQUARE ROOT TIME

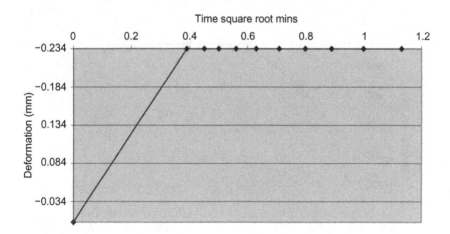

CHANGE IN SPECIMEN THICKNESS VS DISPLACEMENT

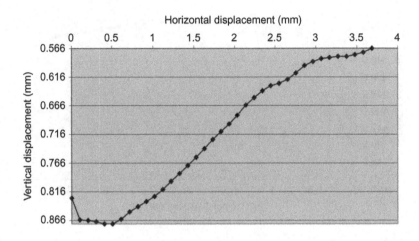

SHEAR STRESS VS DISPLACEMENT

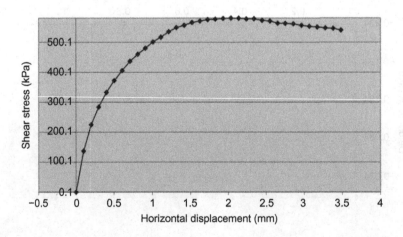

CHANGE IN SPECIMEN THICKNESS VS DISPLACEMENT

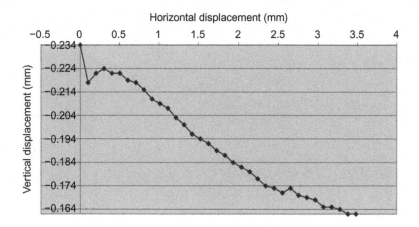

SHEAR STRESS VS DISPLACEMENT

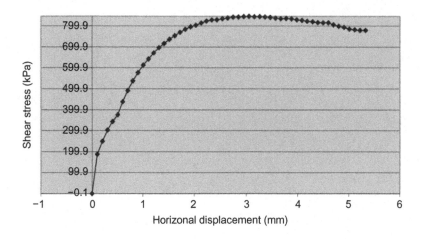

APPENDIX B

Properties of Shear Direct Test in Loose Sand

SETTLEMENT VS SQUARE ROOT TIME

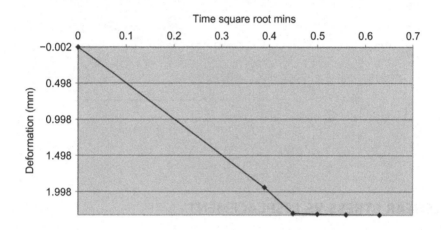

SETTLEMENT VS SQUARE ROOT TIME

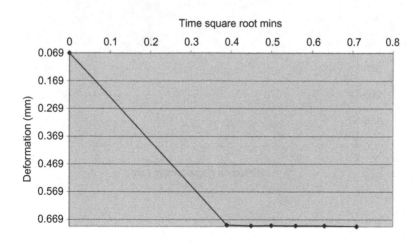

SETTLEMENT VS SQUARE ROOT TIME

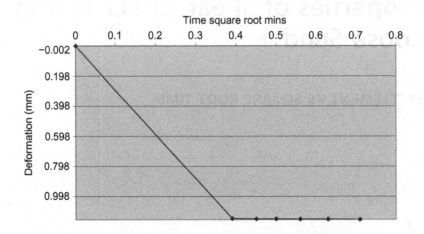

SHEAR STRESS VS DISPLACEMENT

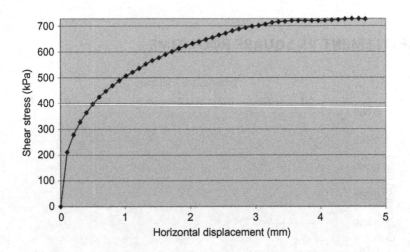

SHEAR STRESS VS DISPLACEMENT

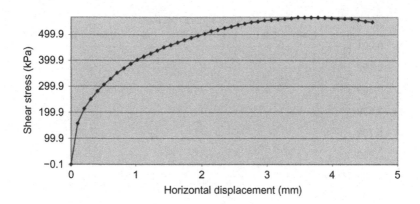

CHANGE IN SPECIMEN THICKNESS VS DISPLACEMENT

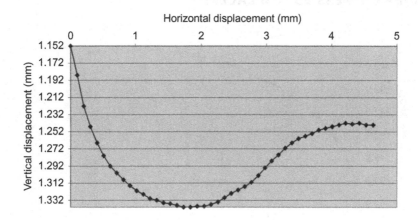

SETTLEMENT VS SQUARE ROOT TIME

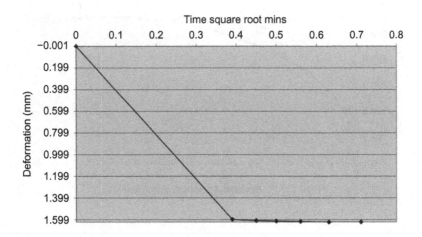

SHEAR STRESS VS DISPLACEMENT

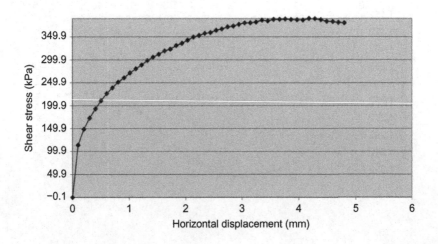

CHANGE IN SPECIMEN THICKNESS VS DISPLACEMENT

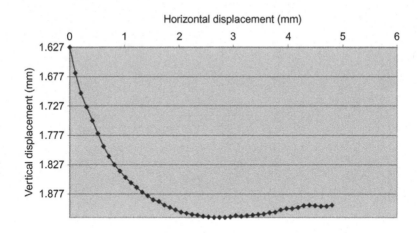

SETTLEMENT VS SQUARE ROOT TIME

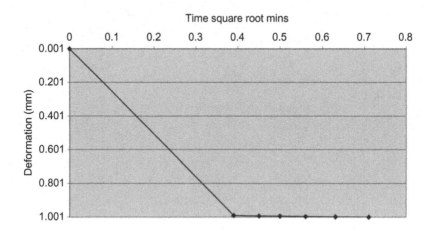

CHANGE IN SPECIMEN THICKNESS VS DISPLACEMENT

SETTLEMENT VS SQUARE ROOT TIME

APPENDIX C

Design and Modeling of Irregular Shape Anchor in CATIA

ANALYSIS 4M

Mesh

Entity	Size
Nodes	3014
Elements	11,349

Element Type

Connectivity	Statistics
TE4	11,349 (100.00%)

Materials.1

Material	Steel
Young modulus	2e + 011N_m2
Poisson ratio	0.266
Density	7860kg_m3
Thermal expansion	1.17e − 005_Kdeg
Yield strength	2.5e + 008N_m2

Static Case
Boundary Conditions

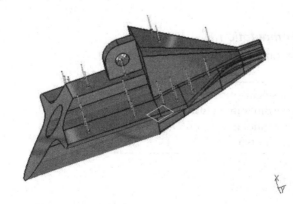

Structure Computation

Number of nodes:	3014
Number of elements:	11,349
Number of D.O.F.:	9042
Number of contact relations:	0
Number of kinematic relations:	0

Linear tetrahedron: 11,349

Restraint Computation
   Name: RestraintSet.1
   Number of S.P.C: 54

Load Computation
   Name: LoadSet.1
   Applied load resultant:
       $Fx = -1.300e+004$ N
       $Fy = 1.281e-007$ N
       $Fz = 7.218e-007$ N
       $Mx = 3.547e-010$ N $\times$ m
       $My = -5.126e+003$ N $\times$ m
       $Mz = -2.226e+001$ N $\times$ m

Stiffness Computation

Number of lines:	9042
Number of coefficients:	165,684
Number of blocks:	1
Maximum number of coefficients per bloc:	165,684
Total matrix size:	1.93 Mb

Singularity Computation
Restraint: RestraintSet.1

Number of local singularities:	0
Number of singularities in translation:	0
Number of singularities in rotation:	0
Generated constraint type:	MPC

Constraint Computation
Restraint: RestraintSet.1

Number of constraints:	54
Number of coefficients:	0
Number of factorized constraints:	54
Number of coefficients:	0
Number of deferred constraints:	0

Factorized Computation

Method:	Sparse
Number of factorized degrees:	8988
Number of super nodes:	1559
Number of overhead indices:	73,944
Number of coefficients:	1,198,437
Maximum front width:	627
Maximum front size:	196,878
Size of the factorized matrix (Mb):	9.14335
Number of blocks:	2
Number of Mflops for factorization:	$3.608e+002$
Number of Mflops for solve:	$4.839e+000$
Minimum relative pivot:	$6.248e-003$

Direct Method Computation
Name: StaticSet.1
Restraint: RestraintSet.1
Load: LoadSet.1
Strain energy: $2.406e-002$ J
Equilibrium

Components	Applied forces	Relative	Reactions	Relative magnitude error
Fx (N)	$-1.3000e+004$	$1.3000e+004$	$1.4588e-009$	$3.3425e-013$
Fy (N)	$1.2806e-007$	$-1.2781e-007$	$2.4821e-010$	$5.6869e-014$
Fz (N)	$7.2177e-007$	$-7.2193e-007$	$-1.5825e-010$	$3.6259e-014$
Mx (N × m)	$3.5473e-010$	$-4.5454e-010$	$-9.9809e-011$	$3.0491e-014$
My (N × m)	$-5.1258e+003$	$5.1258e+003$	$6.2482e-010$	$1.9088e-013$
Mz (N × m)	$-2.2262e+001$	$2.2262e+001$	$-8.4029e-011$	$2.5670e-014$

Static Case Solution.1—Deformed Mesh.2

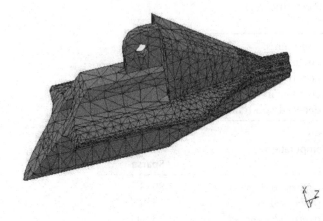

On deformed mesh — On boundary — Over all the model

Static Case Solution.1—Von Misses Stress (nodal values).2

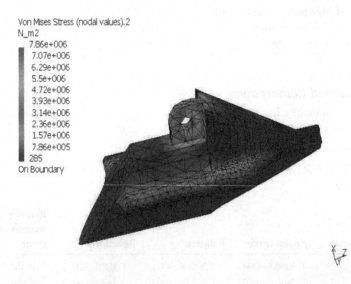

Von Mises Stress (nodal values).2
N_m2
7.86e+006
7.07e+006
6.29e+006
5.5e+006
4.72e+006
3.93e+006
3.14e+006
2.36e+006
1.57e+006
7.86e+005
285
On Boundary

3D elements: Components: All
On deformed mesh — On boundary — Over all the model

Global Sensors

Sensor name	Sensor value
Energy	0.024 J

ANALYSIS 8M

Mesh

Entity	Size
Nodes	3014
Elements	11,349

Element Type

Connectivity	Statistics
TE4	11,349 (100.00%)

Materials.1

Material	Steel
Young modulus	2e + 011N_m2
Poisson ratio	0.266
Density	7860kg_m3
Thermal expansion	1.17e − 005_Kdeg
Yield strength	2.5e + 008N_m2

Static Case
Boundary Conditions

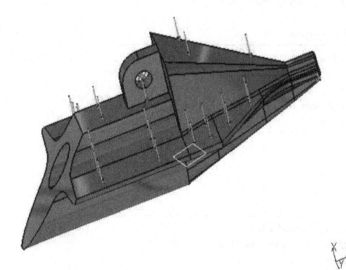

Structure Computation

Number of nodes:	3014
Number of elements:	11,349
Number of D.O.F.:	9042
Number of contact relations:	0
Number of kinematic relations:	0
Linear tetrahedron:	11,349

Restraint Computation
    Name: RestraintSet.1
    Number of S.P.C: 54

Load Computation
    Name: LoadSet.1
    Applied load resultant:
        $Fx = -2.600e + 004$ N
        $Fy = 2.561e - 007$ N
        $Fz = 1.444e - 006$ N
        $Mx = 7.095e - 010$ N × m
        $My = -1.025e + 004$ N × m
        $Mz = -4.452e + 001$ N × m

Stiffness Computation

Number of lines:	9042
Number of coefficients:	165,684
Number of blocks:	1
Maximum number of coefficients per bloc:	165,684
Total matrix size:	1.93 Mb

Singularity Computation
    Restraint: RestraintSet.1

Number of local singularities:	0
Number of singularities in translation:	0
Number of singularities in rotation:	0
Generated constraint type:	MPC

Constraint Computation
Restraint RestraintSet.1

Number of constraints:	54
Number of coefficients:	0
Number of factorized constraints:	54
Number of coefficients:	0
Number of deferred constraints:	0

Factorized Computation

Method:	Sparse
Number of factorized degrees:	8988
Number of super nodes:	1559
Number of overhead indices:	73,944
Number of coefficients:	1,198,437
Maximum front width:	627
Maximum front size:	196,878
Size of the factorized matrix (Mb):	9.14335
Number of blocks:	2
Number of Mflops for factorization:	$3.608e + 002$
Number of Mflops for solve:	$4.839e + 000$
Minimum relative pivot:	$6.248e - 003$

Direct Method Computation
    Name: StaticSet.1
    Restraint: RestraintSet.1
    Load: LoadSet.1
    Strain Energy: $9.625e - 002$ J
Equilibrium

Components	Applied forces	Reactions	Residual	Relative magnitude error
Fx (N)	$-2.6000e + 004$	$2.6000e + 004$	$2.9177e - 009$	$3.3425e - 013$
Fy (N)	$2.5611e - 007$	$-2.5562e - 007$	$4.9641e - 010$	$5.6869e - 014$
Fz (N)	$1.4435e - 006$	$-1.4439e - 006$	$-3.1650e - 010$	$3.6259e - 014$
Mx (N × m)	$7.0946e - 010$	$-9.0908e - 010$	$-1.9962e - 010$	$3.0491e - 014$
My (N × m)	$-1.0252e + 004$	$1.0252e + 004$	$1.2496e - 009$	$1.9088e - 013$
Mz (N × m)	$-4.4524e + 001$	$4.4524e + 001$	$-1.6806e - 010$	$2.5670e - 014$

Static Case Solution.1—Deformed Mesh.2

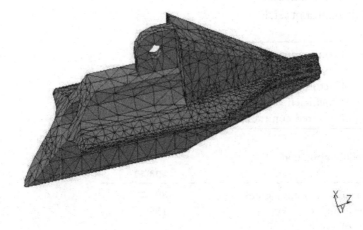

On deformed mesh — On boundary — Over all the model

Static Case Solution.1—Von Mises Stress (nodal values).2

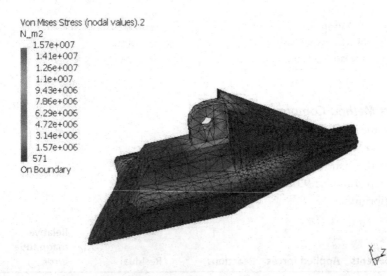

3D elements: Components: All
On deformed mesh — On boundary — Over all the model

Global Sensors	
Sensor name	**Sensor value**
Energy	0.096 J

ANALYSIS 12M

Mesh

Entity	Size
Nodes	3014
Elements	11,349

Element Type

Connectivity	Statistics
TE4	11,349 (100.00%)

Materials.1

Material	Steel
Young modulus	2e + 011N_m2
Poisson ratio	0.266
Density	7860kg_m3
Thermal expansion	1.17e − 005_Kdeg
Yield strength	2.5e + 008N_m2

Static Case
Boundary Conditions

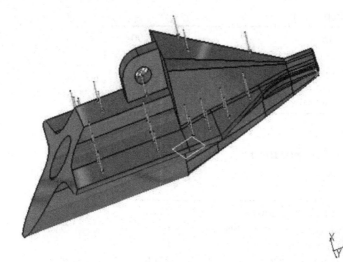

Structure Computation

Number of nodes:	3014
Number of elements:	11,349
Number of D.O.F.:	9042
Number of contact relations:	0
Number of kinematic relations:	0
Linear tetrahedron:	11,349

Restraint Computation
    Name: RestraintSet.1
    Number of S.P.C: 54

Load Computation
    Name: LoadSet.1
    Applied load resultant:
        $Fx = -3.900e + 004$ N
        $Fy = -1.714e - 007$ N
        $Fz = -7.749e - 007$ N
        $Mx = -3.114e - 007$ N × m
        $My = -1.538e + 004$ N × m
        $Mz = -6.679e + 001$ N × m

Stiffness Computation

Number of lines:	9042
Number of coefficients:	165,684
Number of blocks:	1
Maximum number of coefficients per bloc:	165,684
Total matrix size:	1.93 Mb

Singularity Computation

Restraint: RestraintSet.1

Number of local singularities:	0
Number of singularities in translation:	0
Number of singularities in rotation:	0
Generated constraint type:	MPC

Constraint Computation
Restraint: RestraintSet.1

Number of constraints:	54
Number of coefficients:	0
Number of factorized constraints:	54
Number of coefficients:	0
Number of deferred constraints:	0

Factorized Computation

Method:	Sparse
Number of factorized degrees:	8988
Number of super nodes:	1559
Number of overhead indices:	73,944
Number of coefficients:	1,198,437
Maximum front width:	627
Maximum front size:	196,878
Size of the factorized matrix (Mb):	9.14335
Number of blocks:	2
Number of Mflops for factorization:	$3.608e + 002$
Number of Mflops for solve:	$4.839e + 000$
Minimum relative pivot:	$6.248e - 003$

Direct Method Computation
Name: StaticSet.1
Restraint: RestraintSet.1
Load: LoadSet.1
Strain energy: $2.166e - 001$ J
Equilibrium

Components	Applied forces	Reactions	Residual	Relative magnitude error
Fx (N)	$-3.9000e + 004$	$3.9000e + 004$	$4.4092e - 009$	$3.3675e - 013$
Fy (N)	$-1.7136e - 007$	$1.7210e - 007$	$7.3726e - 010$	$5.6307e - 014$
Fz (N)	$-7.7486e - 007$	$7.7437e - 007$	$-4.8749e - 010$	$3.7231e - 014$
Mx (N × m)	$-3.1139e - 007$	$3.1109e - 007$	$-2.9773e - 010$	$3.0318e - 014$
My (N × m)	$-1.5378e + 004$	$1.5378e + 004$	$1.8954e - 009$	$1.9301e - 013$
Mz (N × m)	$-6.6786e + 001$	$6.6786e + 001$	$-2.5332e - 010$	$2.5796e - 014$

Static Case Solution.1—Deformed Mesh.2

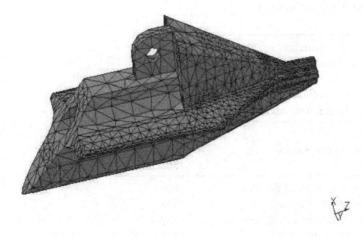

On deformed mesh — On boundary — Over all the model

Static Case Solution.1—Von Mises Stress (nodal values).2

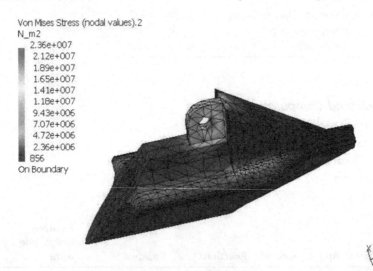

Von Mises Stress (nodal values).2
N_m2
2.36e+007
2.12e+007
1.89e+007
1.65e+007
1.41e+007
1.18e+007
9.43e+006
7.07e+006
4.72e+006
2.36e+006
856
On Boundary

3D elements: Components: All
On deformed mesh — On boundary — Over all the model

Global Sensors	
Sensor name	Sensor value
Energy	0.217 J

ANALYSIS 16M

Mesh

Entity	Size
Nodes	3014
Elements	11,349

Element Type

Connectivity	Statistics
TE4	11,349 (100.00%)

Materials.1

Material	Steel
Young modulus	2e + 011N_m2
Poisson ratio	0.266
Density	7860kg_m3
Thermal expansion	1.17e − 005_Kdeg
Yield strength	2.5e + 008N_m2

Static Case
Boundary Conditions

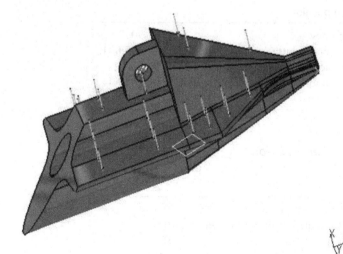

Structure Computation

Number of nodes:	3014
Number of elements:	11,349
Number of D.O.F.	9042
Number of contact relations:	0
Number of kinematic relations:	0
Linear tetrahedron:	11,349

Restraint Computation

Name: RestraintSet.1

Number of S.P.C: 54

Load Computation

Name: LoadSet.1

Applied load resultant:

$Fx = -5.200e + 004$ N

$Fy = 5.122e - 007$ N

$Fz = 2.887e - 006$ N

$Mx = 1.419e - 009$ N $\times$ m

$My = -2.050e + 004$ N $\times$ m

$Mz = -8.905e + 001$ N $\times$ m

Stiffness Computation

Number of lines:	9042
Number of coefficients:	165,684
Number of blocks:	1
Maximum number of coefficients per bloc:	165,684
Total matrix size:	1.93 Mb

Singularity Computation

Restraint: RestraintSet.1

Number of local singularities:	0
Number of singularities in translation:	0
Number of singularities in rotation:	0
Generated constraint type:	MPC

Constraint Computation
Restraint: RestraintSet.1

Number of constraints:	54
Number of coefficients:	0
Number of factorized constraints:	54
Number of coefficients:	0
Number of deferred constraints:	0

Factorized Computation

Method:	Sparse
Number of factorized degrees:	8988
Number of super nodes:	1559
Number of overhead indices:	73,944
Number of coefficients:	1,198,437
Maximum front width:	627
Maximum front size:	196,878
Size of the factorized matrix (Mb):	9.14335
Number of blocks:	2
Number of Mflops for factorization:	$3.608e+002$
Number of Mflops for solve:	$4.839e+000$
Minimum relative pivot:	$6.248e-003$

Direct Method Computation
   Name: StaticSet.1
   Restraint: RestraintSet.1
   Load: LoadSet.1
   Strain energy: $3.850e-001$ J
Equilibrium

Components	Applied forces	Reactions	Residual	Relative magnitude error
Fx (N)	$-5.2000e+004$	$5.2000e+004$	$5.8353e-009$	$3.3425e-013$
Fy (N)	$5.1223e-007$	$-5.1123e-007$	$9.9283e-010$	$5.6869e-014$
Fz (N)	$2.8871e-006$	$-2.8877e-006$	$-6.3301e-010$	$3.6259e-014$
Mx (N × m)	$1.4189e-009$	$-1.8182e-009$	$-3.9924e-010$	$3.0491e-014$
My (N × m)	$-2.0503e+004$	$2.0503e+004$	$2.4993e-009$	$1.9088e-013$
Mz (N × m)	$-8.9049e+001$	$8.9049e+001$	$-3.3612e-010$	$2.5670e-014$

Static Case Solution.1—Deformed Mesh.2

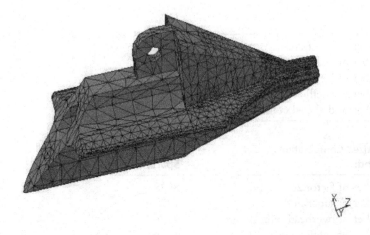

On deformed mesh — On boundary — Over all the model

Static Case Solution.1—Von Mises Stress (nodal values).2

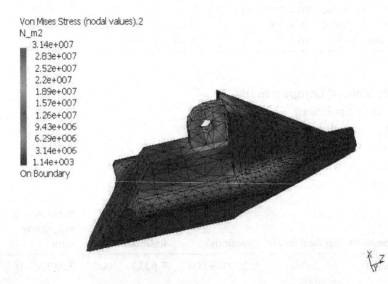

Von Mises Stress (nodal values).2
N_m2
3.14e+007
2.83e+007
2.52e+007
2.2e+007
1.89e+007
1.57e+007
1.26e+007
9.43e+006
6.29e+006
3.14e+006
1.14e+003
On Boundary

3D elements: Components: All
On deformed mesh — On boundary — Over all the model

Global Sensors

Sensor name	Sensor value
Energy	0.385 J

ANALYSIS 20M

Mesh

Entity	Size
Nodes	3014
Elements	11,349

Element Type

Connectivity	Statistics
TE4	11,349 (100.00%)

Materials.1

Material	Steel
Young modulus	2e + 011N_m2
Poisson ratio	0.266
Density	7860kg_m3
Thermal expansion	1.17e − 005_Kdeg
Yield strength	2.5e + 008N_m2

Static Case

Boundary Conditions

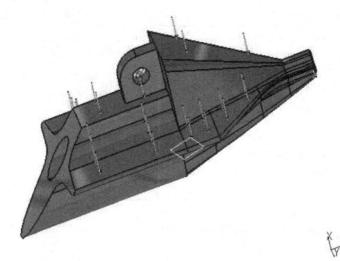

Structure Computation

Number of nodes:	3014
Number of elements:	11,349
Number of D.O.F.:	9042
Number of contact relations:	0
Number of kinematic relations:	0
Linear tetrahedron:	11,349

Restraint Computation

Name: RestraintSet.1

Number of S.P.C: 54

Load Computation

Name: LoadSet.1

Applied load resultant:

$Fx = -6.500e + 004$ N

$Fy = -1.233e - 006$ N

$Fz = 2.190e - 006$ N

$Mx = 4.348e - 007$ N $\times$ m

$My = -2.563e + 004$ N $\times$ m

$Mz = -1.113e + 002$ N $\times$ m

Stiffness Computation

Number of lines:	9042
Number of coefficients:	165,684
Number of blocks:	1
Maximum number of coefficients per bloc:	165,684
Total matrix size:	1.93 Mb

Singularity Computation

Restraint: RestraintSet.1

Number of local singularities:	0
Number of singularities in translation:	0
Number of singularities in rotation:	0
Generated constraint type:	MPC

Constraint Computation
Restraint: RestraintSet.1

Number of constraints:	54
Number of coefficients:	0
Number of factorized constraints:	54
Number of coefficients:	0
Number of deferred constraints:	0

Factorized Computation Method:	Sparse
Number of factorized degrees:	8988
Number of super nodes:	1559
Number of overhead indices:	73,944
Number of coefficients:	1,198,437
Maximum front width:	627
Maximum front size:	196,878
Size of the factorized matrix (Mb):	9.14335
Number of blocks:	2
Number of Mflops for factorization:	$3.608e+002$
Number of Mflops for solve:	$4.839e+000$
Minimum relative pivot:	$6.248e-003$

Direct Method Computation
Name: StaticSet.1
Restraint: RestraintSet.1
Load: LoadSet.1
Strain energy: $6.016e-001$ J
Equilibrium

Components	Applied forces	Reactions	Residual	Relative magnitude error
Fx (N)	$-6.5000e+004$	$6.5000e+004$	$7.2614e-009$	$3.3275e-013$
Fy (N)	$-1.2331e-006$	$1.2343e-006$	$1.1917e-009$	$5.4607e-014$
Fz (N)	$2.1905e-006$	$-2.1913e-006$	$-8.6038e-010$	$3.9426e-014$
Mx (N × m)	$4.3477e-007$	$-4.3525e-007$	$-4.8546e-010$	$2.9661e-014$
My (N × m)	$-2.5629e+004$	$2.5629e+004$	$3.1650e-009$	$1.9338e-013$
Mz (N × m)	$-1.1131e+002$	$1.1131e+002$	$-4.2181e-010$	$2.5772e-014$

Static Case Solution.1—Deformed Mesh.2

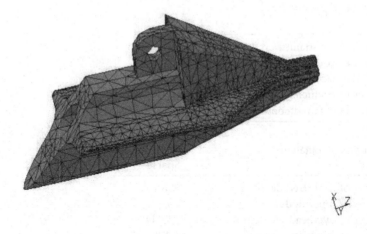

On deformed mesh — On boundary — Over all the model

Static Case Solution.1—Von Mises Stress (nodal values).2

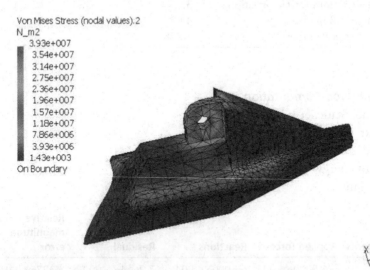

Von Mises Stress (nodal values).2
N_m2
3.93e+007
3.54e+007
3.14e+007
2.75e+007
2.36e+007
1.96e+007
1.57e+007
1.18e+007
7.86e+006
3.93e+006
1.43e+003
On Boundary

3D elements: Components: All

On deformed mesh — On boundary — Over all the model

Global Sensors

Sensor name	Sensor value
Energy	0.602 J

ANALYSIS 40M

Mesh

Entity	Size
Nodes	3014
Elements	11,349

Element Type

Connectivity	Statistics
TE4	11,349 (100.00%)

Materials.1

Material	Steel
Young modulus	2e + 011N_m2
Poisson ratio	0.266
Density	7860kg_m3
Thermal expansion	1.17e − 005_Kdeg
Yield strength	2.5e + 008N_m2

Static Case
Boundary Conditions

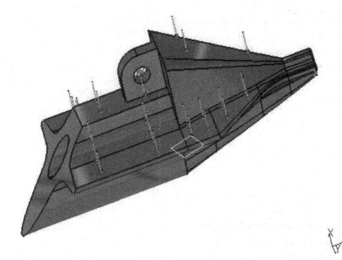

Structure Computation

Number of nodes:	3014
Number of elements:	11,349
Number of D.O.F.:	9042
Number of contact relations:	0
Number of kinematic relations:	0
Linear tetrahedron:	11,349

Restraint Computation

Name: RestraintSet.1

Number of S.P.C: 54

Load Computation

Name: LoadSet.1

Applied load resultant:

$Fx = -1.300e + 005 \ N$

$Fy = -2.466e - 006 \ N$

$Fz = 4.381e - 006 \ N$

$Mx = 8.695e - 007 \ N \times m$

$My = -5.126e + 004 \ N \times m$

$Mz = -2.226e + 002 \ N \times m$

Stiffness Computation

Number of lines:	9042
Number of coefficients:	165,684
Number of blocks:	1
Maximum number of coefficients per bloc:	165,684
Total matrix size:	1.93 Mb

Singularity Computation

Restraint: RestraintSet.1

Number of local singularities:	0
Number of singularities in translation:	0
Number of singularities in rotation:	0
Generated constraint type:	MPC

Constraint Computation

Restraint: RestraintSet.1

Number of constraints:	54
Number of coefficients:	0
Number of factorized constraints:	54
Number of coefficients:	0
Number of deferred constraints:	0

Factorized Computation

Method:	Sparse
Number of factorized degrees:	8988
Number of super nodes:	1559
Number of overhead indices:	73,944
Number of coefficients:	1,198,437
Maximum front width:	627
Maximum front size:	196,878
Size of the factorized matrix (Mb):	9.14335
Number of blocks:	2
Number of Mflops for factorization:	$3.608e+002$
Number of Mflops for solve:	$4.839e+000$
Minimum relative pivot:	$6.248e-003$

Direct Method Computation

Name: StaticSet.1
Restraint: RestraintSet.1
Load: LoadSet.1
Strain energy: $2.406e+000$ J

Equilibrium

Components	Applied forces	Reactions	Residual	Relative magnitude error
Fx (N)	$-1.3000e+005$	$1.3000e+005$	$1.4581e-008$	$3.3408e-013$
Fy (N)	$-2.4661e-006$	$2.4685e-006$	$2.3929e-009$	$5.4825e-014$
Fz (N)	$4.3809e-006$	$-4.3827e-006$	$-1.7462e-009$	$4.0009e-014$
Mx (N × m)	$8.6953e-007$	$-8.7051e-007$	$-9.7484e-010$	$2.9781e-014$
My (N × m)	$-5.1258e+004$	$5.1258e+004$	$6.3446e-009$	$1.9382e-013$
Mz (N × m)	$-2.2262e+002$	$2.2262e+002$	$-8.4231e-010$	$2.5732e-014$

Static Case Solution.1—Deformed Mesh.2

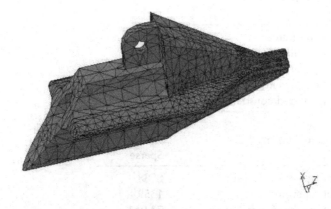

On deformed mesh — On boundary — Over all the model

Static Case Solution.1—Von Mises Stress (nodal values).2

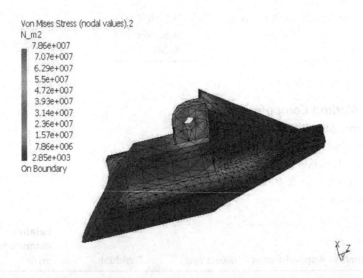

3D elements: Components: All
On deformed mesh — On boundary — Over all the model

Global Sensors

Sensor name	Sensor value
Energy	2.406 J

Impact Test

ANALYSIS 25000N

Mesh

Entity	Size
Nodes	19,357
Elements	11,305

Element Type

Connectivity	Statistics
TE10	11,305 (100.00%)

Materials.1

Material	Steel
Young modulus	2e + 011N_m2
Poisson ratio	0.266
Density	7860kg_m3
Thermal expansion	1.17e − 005_Kdeg
Yield strength	2.5e + 008N_m2

Static Case
Boundary Conditions

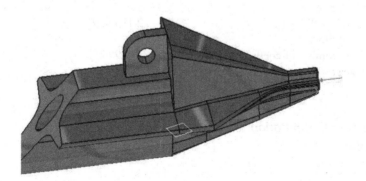

Structure Computation

Number of nodes:	19,357
Number of elements:	11,305
Number of D.O.F.:	58,071
Number of Contact relations:	0
Number of Kinematic relations:	0
Parabolic tetrahedron:	11,305

Restraint Computation
    Name: RestraintSet.1
    Number of S.P.C: 75

Load Computation
    Name: LoadSet.1
    Applied load resultant:
        $Fx = -3.260e - 009$ N
        $Fy = 1.208e - 009$ N
        $Fz = -2.500e + 004$ N
        $Mx = 6.339e + 000$ N × m
        $My = -1.285e + 002$ N × m
        $Mz = -8.762e - 011$ N × m

Stiffness Computation

Number of lines:	58,071
Number of coefficients:	2,193,819
Number of blocks:	5
Maximum number of coefficients per bloc:	499,990
Total matrix size:	25.33 Mb

Singularity Computation
Restraint: RestraintSet.1

Number of local singularities:	0
Number of singularities in translation:	0
Number of singularities in rotation:	0
Generated constraint type:	MPC

Constraint Computation
Restraint: RestraintSet.1

Number of constraints:	75
Number of coefficients:	0
Number of factorized constraints:	75
Number of coefficients:	0
Number of deferred constraints:	0

Factorized Computation

Method:	Sparse
Number of factorized degrees:	57,996
Number of supernodes:	4187
Number of overhead indices:	505,809
Number of coefficients:	19,813,896
Maximum front width:	1914
Maximum front size:	1,832,655
Size of the factorized matrix (Mb):	151.168
Number of blocks:	20
Number of Mflops for factorization:	$1.523e + 004$
Number of Mflops for solve:	$7.955e + 001$
Minimum relative pivot:	$3.895e - 002$

Direct Method Computation
Name: StaticSet.1
Restraint: RestraintSet.1
Load: LoadSet.1
Strain energy: $1.149e - 001$ J
Equilibrium

Components	Applied forces	Reactions	Residual	Relative magnitude error
Fx (N)	$-3.2596e - 009$	$3.1400e - 009$	$-1.1958e - 010$	$5.2986e - 014$
Fy (N)	$1.2078e - 009$	$-1.1561e - 009$	$5.1720e - 011$	$2.2917e - 014$
Fz (N)	$-2.5000e + 004$	$2.5000e + 004$	$-7.1668e - 010$	$3.1755e - 013$
Mx (N × m)	$6.3386e + 000$	$-6.3386e + 000$	$-2.5035e - 011$	$1.4790e - 014$
My (N × m)	$-1.2852e + 002$	$1.2852e + 002$	$-7.3641e - 011$	$4.3506e - 014$
Mz (N × m)	$-8.7619e - 011$	$9.6196e - 011$	$8.5767e - 012$	$5.0670e - 015$

Static Case Solution.1—Deformed Mesh.1

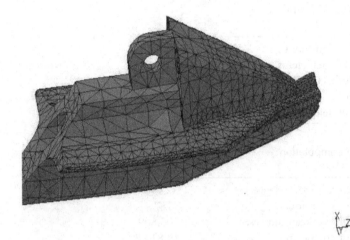

On deformed mesh — On boundary — Over all the model

Static Case Solution.1—Von Mises Stress (nodal values).2

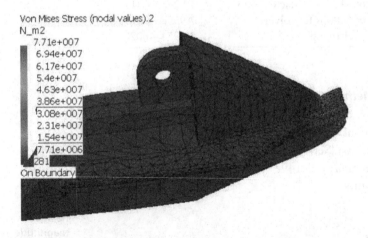

3D elements: Components: All
On deformed mesh — On boundary — Over all the model

Global Sensors

Sensor name	Sensor value
Energy	0.115 J

APPENDIX D

Variation of Breakout Factor With Embedment Ratio Based on Author's Empirical Formula

Models	Loose sand		Dense sand	
	L/D	Breakout factor	L/D	Breakout factor
Small model	1	0.806	1	4.31
	2	2.453	2	8.73
	3	3.538	3	12.68
	4	4.061	4	16.17
Big model	1	0.806	1	4.31
	2	2.453	2	8.73
	3	3.536	3	12.68
	4	4.061	4	16.17

Author's empirical formula in loose sand: $N_q = -0.281\left(\dfrac{L}{D}\right)^2 + 2.49\left(\dfrac{L}{D}\right) - 1.403$

Author's empirical formula in dense sand: $N_q = -0.233\left(\dfrac{L}{D}\right)^2 + 5.119\left(\dfrac{L}{D}\right) - 0.576$

Friction angle in loose sand: $\varphi = 35°$
Friction angle in dense sand: $\varphi = 42°$

APPENDIX E

Variation of Breakout Factor With Embedment Ratio Based on Balla's Theory (1961) in Loose Sand

[1]	[2]	[3]	[4]	[5]	[6]	[7]	[8]	[9]	[10]
Model size	D (mm)	L/D	Depth (mm)	$F_1 + F_3$	Theoretical pullout load (N) in circular	Breakout factor, N_q in circular	Breakout factor, N_q in square	Breakout factor, N_q in strip	Theoretical pullout load, P_u (N)
Small model	159	1	159	2.37	141	3.01	2.76	1.42	33.76
		2	318	1.19	570	6.06	5.31	2.01	95.82
		3	477	0.88	1423	10.15	8.67	3.06	217.85
		4	636	0.72	2759	14.68	12.22	4.14	392.98
Big size	297	1	297	2.37	925	3.03	2.77	1.45	169.42
		2	594	1.19	3716	6.10	5.34	2.03	474.38
		3	891	0.88	9274	10.12	8.65	3.2	1121.68
		4	1188	0.72	17987	15.70	13.07	4.38	2047.08

Friction angle in loose sand: $\varphi = 35°$

Unit weight: 14.90 kN/m³

$F_1 + F_3$: Fig. 2.10

Theoretical pullout load (N) in circular: $\gamma(F_1 + F_3)H^3$

Breakout factor, N_q in circular: $\dfrac{P_u}{\gamma AH}$

Breakout factor, N_q in square: $N_q(\text{square}) * \text{correlation factor}$

Breakout factor, N_q in strip: $\dfrac{N_q(\text{Square})}{\left(\dfrac{10}{3} - \dfrac{1.4}{L/D}\right)}$

Theoretical pullout load, P_u (N) in strip: γAHN_q

Variation of Breakout Factor With Embedment Ratio Based on Balla's Theory (1961) in Dense Sand

[1] Model size	[2] D (mm)	[3] L/D	[4] Depth (mm)	[5] F_1+F_3	[6] Theoretical pullout load (N) in circular	[7] Breakout factor, N_q in circular	[8] Breakout factor, N_q in square	[9] Breakout factor, N_q in strip	[10] Theoretical pullout load, P_u (N)
Small model	159	1	159	2.45	166.92	3.12	2.86	1.50	40.62
		2	318	1.26	686.78	6.42	5.63	2.14	115.54
		3	477	0.91	1674.04	10.43	8.91	3.14	254.30
		4	636	0.77	3357.62	15.69	13.06	4.38	472.97
Big size	297	1	297	2.45	1087.94	3.14	2.88	1.51	200.7
		2	594	1.26	4476.09	6.46	5.66	2.15	572.07
		3	891	0.91	10910.48	10.50	8.97	3.17	1264.84
		4	1188	0.77	21883.14	15.80	13.16	4.41	2344.67

Friction angle in dense sand: $\varphi = 42°$

Unit weight: 16.95 kN/m³

F_1+F_3: Fig. 2.10

Theoretical pullout load (N) in circular: $\gamma(F_1+F_3)H^3$

Breakout factor, N_q in circular: $\dfrac{P_u}{\gamma AH}$

Breakout factor, N_q in square: $N_q(\text{square}) * \text{correlation factor}$

Breakout factor, N_q in strip: $\dfrac{N_q(\text{square})}{\frac{10}{3} - \frac{1.4}{L/D}}$

Theoretical pullout load, P_u (N) in strip: γAHN_q

APPENDIX G

Variation of Breakout Factor With Embedment Ratio Based on Meyerhof and Adams Theory (1968) in Loose Sand

[1]	[2]	[3]	[4]	[5]	[6]	[7]	[8]
Model size	D (mm)	L/D	Depth (mm)	Shape factor	K_u $\tan\varphi$	Theoretical pullout load, P_u (N)	Breakout factor, N_q
Small	159	1	159	1.25	0.65	84.94	3.57
model		2	318	1.5	0.65	321.68	6.77
		3	477	1.75	0.65	768.63	10.79
		4	636	2	0.65	1458.22	15.36
Big size	297	1	297	1.25	0.65	484.48	4.14
		2	594	1.5	0.65	1856.17	7.94
		3	891	1.75	0.65	4342.89	12.38
		4	1188	2	0.65	8172.51	17.18

Friction angle in loose sand: $\varphi = 35°$.
Unit weight: 14.90 kN/m^3.
Shape factor: m was showed Fig. 2.12.

$$S_f = 1 + m\frac{L}{D}$$

K_u: Fig. 2.11
Theoretical pullout load, P_u (N):

$$P_u = W + \gamma H^2 \left(2S_f L + B - L\right)K_u \tan\varphi$$

Breakout factor, N_q:

$$N_q = 1 + \frac{L}{D_u}\tan\varphi$$

Variation of Breakout Factor With Embedment Ratio Based on Meyerhof and Adams Theory (1968) in Loose Sand

APPENDIX H

Variation of Breakout Factor With Embedment Ratio Based on Meyerhof and Adams Theory (1968) in Dense Sand

[1]	[2]	[3]	[4]	[5]	[6]	[7]	[8]
Model size	D (mm)	L/D	Depth (mm)	Shape factor	K_u $\tan\varphi$	Theoretical pullout load, P_u (N)	Breakout factor, N_q
Small model	159	1	159	1.39	0.864	127.37	4.71
		2	318	1.78	0.864	528.30	9.78
		3	477	2.17	0.864	1311.93	16.20
		4	636	2.56	0.864	2587.43	23.96
Big size	297	1	297	1.39	0.864	720.99	5.42
		2	594	1.78	0.864	2976.46	11.19
		3	891	2.17	0.864	7303.86	18.31
		4	1188	2.56	0.864	14240.65	26.78

Friction angle in dense sand: $\varphi = 42°$
Unit weight: 16.95 kN/m^3
Shape factor: m was showed Fig. 2.12

$$S_f = 1 + m\frac{L}{D}$$

K_u: Fig. 2.11
Theoretical pullout load, P_u (N):

$$P_u = W + \gamma H^2 \left(2S_f L + B - L\right) K_u \tan\varphi$$

Breakout factor, N_q:

$$N_q = 1 + \frac{L}{D} K_u \tan\varphi$$

Variation of Breakout Factor With Embedment Ratio Based on Meyerhof and Adams Theory (1968) in Dense Sand

Model size	D (mm)	L/D	Depth (mm)	Model qu (kPa)	K p	Embedment ratio	Theoretical pullout load (N)	Breakout factor, N_q

APPENDIX I

Variation of Breakout Factor With Embedment Ratio Based on Vesic Theory (1971) in Loose Sand

[1]	[2]	[3]	[4]	[5]	[6]
Model size	D (mm)	L/D	Depth (mm)	Breakout factor, N_q	Theoretical pullout load, P_u (N)
Small model	159	1	159	1.55	36.78
		2	318	2.12	100.62
		3	477	2.54	180.83
		4	636	3.15	299
Big size	297	1	297	1.55	181.10
		2	594	2.12	495.41
		3	891	2.54	890.34
		4	1188	3.15	1472.21

Friction angle in loose sand: $\varphi = 35°$
Unit weight: 14.90 kN/m^3
Breakout factor, N_q:

$$N_q = \left[1 + A_1 \left(\frac{H}{\frac{h_1}{2}} \right) + A_2 \left(\frac{H}{\frac{h_1}{2}} \right)^2 \right]$$

Theoretical pullout load, P_u (N):

$$P_u = \gamma H A N_q$$

Variation of Breakout Factor With Embedment Ratio Based on Vesic Theory (1971) in Loose Sand

APPENDIX J

Variation of Breakout Factor With Embedment Ratio Based on Vesic Theory (1971) in Dense Sand

[1]	[2]	[3]	[4]	[5]	[6]
Model size	D (mm)	L/D	Depth (mm)	Breakout factor, N_q	Theoretical pullout load, P_u (N)
Small model	159	1	159	1.6	43.19
		2	318	2.20	118.78
		3	477	2.81	227.57
		4	636	3.40	367.14
Big size	297	1	297	1.6	212.66
		2	594	2.20	584.83
		3	891	2.81	1120.5
		4	1188	3.40	1807.68

Friction angle in dense sand: $\varphi = 42°$
Unit weight: 16.95 kN/m^3
Breakout factor, N_q:

$$N_q = \left[1 + A_1 \left(\frac{H}{\frac{h_1}{2}} \right) + A_2 \left(\frac{H}{\frac{h_1}{2}} \right)^2 \right]$$

Theoretical pullout load, P_u (N):

$$P_u = \gamma H A N_q$$

APPENDIX K

Variation of Breakout Factor With Embedment Ratio Based on Rowe Davis Theory (1982) in Loose Sand

[1]	[2]	[3]	[4]	[5]
Model size	D (mm)	L/D	Depth (mm)	Breakout factor, N_q
Small model	159	1	159	1.28
		2	318	2.90
		3	477	4.20
		4	636	5.98
Big size	297	1	297	1.28
		2	594	2.90
		3	891	4.20
		4	1188	5.98

Friction angle in loose sand: $\varphi = 35°$
Unit weight: 14.90 kN/m^3
Breakout factor, N_q: Fig. 2.16

APPENDIX L

Variation of Breakout Factor With Embedment Ratio Based on Rowe Davis Theory (1982) in Dense Sand

[1] Model size	[2] D (mm)	[3] L/D	[4] Depth (mm)	[5] Breakout factor, N_q
Small model	159	1	159	1.32
		2	318	3.01
		3	477	4.05
		4	636	5.76
Big size	297	1	297	1.32
		2	594	3.01
		3	891	4.05
		4	1188	5.76

Friction angle in dense sand: $\varphi = 42°$
Unit weight: 16.95 kN/m³
Breakout factor, N_q: Fig. 2.16

APPENDIX M

Variation of Breakout Factor With Embedment Ratio Based on Murray and Geddes Theory (1987) in Loose Sand

[1]	[2]	[3]	[4]	[5]
Model size	D (mm)	L/D	Depth (mm)	Breakout factor, N_q
Small model	159	1	159	5.95
		2	318	13.27
		3	477	24.92
		4	636	40.25
Big Size	297	1	297	6.44
		2	594	16.70
		3	891	31.78
		4	1188	51.67

Friction angle in loose sand: $\varphi = 35°$
Unit weight: 14.90 kN/m^3
Breakout factor, N_q:

$$N_q = 1 + \frac{L}{D}\tan\varphi\left(1 + \frac{D}{B} + \frac{\pi L}{3 B}\right)$$

APPENDIX N

Variation of Breakout Factor With Embedment Ratio Based on Murray and Geddes Theory (1987) in Dense Sand

[1]	[2]	[3]	[4]	[5]
Model size	D (mm)	L/D	Depth (mm)	Breakout factor, N_q
Small model	159	1	159	6.53
		2	318	16.78
		3	477	31.76
		4	636	51.46
Big size	297	1	297	8
		2	594	21.19
		3	891	40.57
		4	1188	66.09

Friction angle in dense sand: $\varphi = 42°$
Unit weight: 16.95 kN/m³
Breakout factor, N_q:

$$N_q = 1 + \frac{L}{D}\tan\varphi\left(1 + \frac{D}{B} + \frac{\pi L}{3B}\right)$$

APPENDIX O

Variation of Breakout Factor With Embedment Ratio Based on Dickin and Laman Finding (2007) in Loose Sand

[1]	[2]	[3]	[4]	[5]
Model size	D (mm)	L/D	Depth (mm)	Breakout factor, N_q
Small model	159	1	159	1.24
		2	318	1.91
		3	477	2.42
		4	636	2.98
Big size	297	1	297	1.94
		2	594	2.96
		3	891	3.97
		4	1188	4.42

Friction angle in loose sand: $\varphi = 35°$
Unit weight: 14.90 kN/m^3
Breakout factor, N_q: Fig. 2.22

APPENDIX P

Variation of Breakout Factor With Embedment Ratio Based on Dickin and Laman Finding (2007) in Dense Sand

[1]	[2]	[3]	[4]	[5]
Model size	D (mm)	L/D	Depth (mm)	Breakout factor, N_q
Small model	159	1	159	1.24
		2	318	1.91
		3	477	2.42
		4	636	2.98
Big size	297	1	297	1.94
		2	594	2.96
		3	891	3.97
		4	1188	4.42

Friction angle in dense sand: $\varphi = 42°$
Unit weight: 16.95 kN/m^3
Breakout factor, N_q: Fig. 2.23

INDEX

Note: Page numbers followed by *f* indicate figures, and *t* indicate tables.

Printed in the United States
By Bookmasters